# DISCRETE
# STRUCTURES

(B.Sc. Computer Science Honors NEP Syllabus)

# DISCRETE
# STRUCTURES

(B.Sc. Computer Science Honors NEP Syllabus)


Dr. G. Sudhamathy

Assistant Professor (SS),

Department of Computer Science,

Avinashilingam Institute for Home Science and Higher Education for Women, Coimbatore, India

ISBN 9789355286536  **MJP Publishers**

All rights reserved  No. 44, Nallathambi Street,
Printed and bound in India  Triplicane, Chennai 600 005

MJP 1682  © Publishers, 2024

Publisher :  C. Janarthanan

# FOREWORD

In the dynamic landscape of computer science education, a solid grasp of Discrete Structures is fundamental. As computer science students journey through their B.Sc. Honors program, navigating the intricate paths of the National Education Policy (NEP) syllabus, this book stands as a beacon, illuminating the abstract concepts and practical applications of discrete mathematics.

Aligned meticulously with the NEP syllabus, this comprehensive guide serves as a trusted companion, offering clarity and depth in equal measure. Through its pages, students will embark on a journey of exploration and understanding, unraveling the mysteries of sets, relations, functions, and beyond

As the digital age unfolds, the importance of discrete mathematics in shaping the computational landscape cannot be overstated. It is our hope that this book will not only aid students in mastering the intricacies of their curriculum but also inspire a lifelong passion for mathematical inquiry and problem-solving.

Wishing every reader a rewarding and enlightening journey through the realm of Discrete Structures.

**Dr. K. Vasudevan**

M.Sc., M.Phil, Ph.D., (IIT Madras),

Associate Professor,

Department of Mathematics,

Presidency College (Autonomous),

Chennai 600005, Tamil Nadu, India

# PREFACE

Welcome to the realm of Discrete Structures—a foundational pillar of the B.Sc. Computer Science Honors program aligned with the National Education Policy (NEP). As computing technologies continue to permeate every facet of our lives, the need for a rigorous understanding of fundamental mathematical concepts has never been more pressing. This book is meticulously crafted to cater to the evolving needs of aspiring computer scientists, providing a comprehensive and pedagogically sound exploration of discrete mathematics.

Designed in accordance with the NEP syllabus, this book serves as a roadmap for students navigating the intricacies of discrete structures. Whether you are embarking on your academic journey or seeking to enhance your computational prowess, the content presented herein is carefully curated to align with the learning objectives outlined by the NEP, ensuring a seamless integration into your curriculum.

Each chapter is thoughtfully structured to facilitate a progressive learning experience, starting from the basics and gradually building up to more advanced topics. Emphasizing conceptual clarity and practical relevance, we delve into fundamental concepts such as sets, relations, functions, and logic, laying a solid foundation upon which the edifice of discrete mathematics is erected.

Furthermore, special attention is devoted to topics of paramount importance in computer science, including combinatorics, graph theory, and recurrence relations, illuminating their role in the design and analysis of algorithms, data structures, and computational models. Through a judicious blend of theory, examples, and exercises, we aim to equip you with the analytical tools and problem-solving skills necessary to thrive in today's dynamic and competitive landscape.

In addition to catering to the academic needs of students, this book also serves as a valuable resource for educators and practitioners alike. With its comprehensive coverage, lucid explanations, and abundant examples, it can be seamlessly integrated into classroom instruction, self-study regimens, and technical training programs, fostering a culture of mathematical excellence and computational literacy.

As author, my primary goal is to demystify the enigmatic realm of discrete mathematics and instill in you a sense of intellectual curiosity and mathematical rigor. We invite you to embark on this intellectual voyage with an open mind and a thirst for knowledge, confident in the belief that the insights gleaned from these pages will serve as a springboard for your academic and professional aspirations.

In closing, we extend our heartfelt gratitude to the educators, students, and practitioners whose invaluable feedback and support have shaped this endeavor. It is our fervent hope that this book will not only serve as a trusted companion on your academic journey but also inspire you to explore the boundless horizons of mathematics and computer science.

**Happy exploring!**

# ACKNOWLEDGEMENT

The author Dr. G. Sudhamathy would like to thank all the authorities of Avinashilingam Institute for Home Science and Higher Education for Women, Coimbatore for providing the opportunity to work in this esteemed institution and support for making this book a reality.

I am grateful to the students and teacher community who kept me on my toes with their constant bombardment of queries which prompted me to learn more, simplify my learning and findings and place them neatly in a book.

I wish to thank all the faculty members of the Department of Computer Science, Avinashilingam Institute for Home Science and Higher Education for Women, Coimbatore for their continuous support and suggestions for this book.

My Special regards for the expert Dr. K. Vasudevan, Chennai who gave his expert opinion in shaping this book into a more appealing format.

Most importantly I would like to thank my family members without whose support this book would not have been a reality.

Last, but not the least, this work is a dedication to God, the Almighty whose grace has showered upon me in making my dream come true.

**G. Sudhamathy**

# CONTENTS

# Chapter 1

# SETS

## 1.1. FINITE AND INFINITE SETS

### Definition of Finite set

Finite sets consist of a limited, countable number of elements that can be exhaustively listed. Countable sets, on the other hand, encompass both finite sets and certain infinite sets. A countable set can be placed in a one-to-one correspondence with the natural numbers, meaning its elements can be arranged in a sequence where each element is uniquely associated with a natural number. While all finite sets are countable due to their finite nature, not all countable sets are finite. For instance, the set of all integers is countable, despite its infinite size, because its elements can be systematically listed.

### Examples of finite sets:

P = { 0, 4, 8, 12, ..., 98}

Q = { a : a is an integer, 2 < a < 100}

A set of all English Alphabets.

### Another example of a Finite set:

Consider the set of all months of a year.

Y = {January, February, March, April, May, June, July, August, September, October, November, December}

n (Y) = 12

This set is considered finite because its number of elements is countable.

## Cardinality of Finite Set

If we use 'a' to represent the number of elements in set A, then cardinality of finite set is given by n(A) = a. This indicates that the cardinality of a finite set directly corresponds to its element count. The cardinality of a finite set is always a natural number, or it may be zero if the set is empty.

The cardinality of set B, consisting of all English alphabets, is 26 because it contains 26 elements, each representing a letter of the alphabet. Hence, denoted as n(B) = 26.

Similarly, a set containing the months in a year would have a cardinality of 12, as there are 12 months in total.

These sets can be listed using curly braces or roster form, which allows us to enumerate all their elements.

## Properties of finite sets:

### Cardinality:

A finite set has a definite count of elements, represented by a natural number or zero. This count is known as the cardinality of the set.

### Countability:

Every element in a finite set can be counted or enumerated. There is a finite sequence that exhaustively lists all the elements of the set.

### Subset and Superset:

Any subset of a finite set is also finite. Conversely, any superset of a finite set can be finite or infinite.

**Operations:**

Union, intersection, and set difference operations on finite sets can result in either finite or infinite sets, depending on the sets involved.

**Power Set:**

The power set of a finite set, which consists of all possible subsets of that set, is finite. The number of elements in the power set grows exponentially with the number of elements in the original set.

**Equality:**

Two finite sets are considered equal if and only if they contain the same elements, regardless of the order or multiplicity of elements.

**Representation:**

Finite sets can be represented using roster notation, where the elements are listed within curly braces { } separated by commas.

**For example:**

P = {1, 2, 3, 5}

Q = {2, 4, 6, 8}

R = {2, 3}

- In the given sets P, Q, and R, all elements are finite and countable.

- Set R is a subset of set P (R $\subset$ P) because all elements of R are present in P, demonstrating that a subset of a finite set is always finite.

- The union of sets P and Q (P U Q) yields {1, 2, 3, 4, 5, 6, 8}, which is also finite.

The number of elements in a power set is calculated as $2^n$, where n is the number of elements in the original set. For instance, if set P contains 4 elements, its power set would consist of $2^4 = 16$ elements.

A non-empty finite set refers to a set with a known or determinable number of elements, often represented by a natural number.

**For example:**

S = {a set representing the number of people living in India}

Though it may be challenging to precisely calculate the number of people in India, it is understood to be a natural number, thus classifying S as a non-empty finite set.

In the context where N represents a set containing natural numbers less than n, its cardinality is indeed n. This set can be represented as N = {1, 2, 3, ..., n}, encompassing all natural numbers up to n.

Regarding the question of whether an empty set is finite, let's clarify. An empty set, denoted as {} or $\phi$, contains no elements. Since a finite set is defined by having a countable number of elements, and an empty set indeed has zero elements, it is considered finite. Hence, an empty set is a finite set.

Moving on to infinite sets, they are characterized by having an uncountable number of elements and cannot be represented in roster form. Instead, they are often denoted by using an ellipsis (...) to signify the infinite nature of the set.

**Examples of infinite sets include:**

The set of all whole numbers: W = {0, 1, 2, 3, 4, ...}

The set of all points on a line

The set of all integers

The cardinality of an infinite set is expressed as $n(A) = \infty$, indicating an unlimited number of elements within the set, unlike finite sets where the cardinality is represented by a specific natural number.

## Properties of infinite sets:

### Union of Sets:

When two infinite sets are combined (unioned), the resulting set is always infinite. This is because even if one of the sets has an infinite number of elements and the other a finite number, the infinite set dominates the cardinality of the union.

### Power Set:

The power set of an infinite set, which consists of all possible subsets of that set, is always infinite. This is because the number of subsets grows exponentially with the number of elements in the original set, leading to an uncountable number of elements in the power set.

### Superset:

Any superset of an infinite set is also infinite. This is intuitive since adding more elements or including the original set itself in a larger set expands the cardinality to infinity.

These properties showcase the unique characteristics of infinite sets and highlight how they differ from finite sets in terms of their behavior under operations like union, power set formation, and containment within larger sets.

## Comparison between finite and infinite sets:

### Element Count:

Finite sets have a countable number of elements, meaning that they can be enumerated and listed in a finite sequence.

Infinite sets have an uncountable number of elements, making it impossible to list or enumerate them in a finite sequence.

### Cardinality:

The cardinality of a finite set is a specific natural number or zero, representing the count of its elements.

The cardinality of an infinite set is often denoted as $\infty$, indicating an unlimited number of elements.

### Equality:

Finite sets can be considered equal if they contain the same elements, allowing for a direct comparison.

Comparing the equality of infinite sets solely based on their elements is insufficient due to their potentially uncountable nature. Instead, equality is determined by the existence of a bijection, a one-to-one correspondence, between their elements.

### Operations:

Union, intersection, and set difference operations on finite sets can result in either finite or infinite sets, depending on the sets involved.

The union, intersection, and set difference of two infinite sets are always infinite.

**Power Set:**

The power set of a finite set is finite, consisting of all possible subsets of that set.

The power set of an infinite set is always infinite due to the exponential growth of subsets with the increase in elements.

**Supersets:**

Any superset of a finite set can be finite or infinite.

Supersets of infinite sets are always infinite.

## To know if a Set is Finite or Infinite:

Determining whether a set is finite or infinite depends on its characteristics and properties. Here are some guidelines to help you identify whether a set is finite or infinite:

**Countability:**

If the set contains a limited, countable number of elements that can be enumerated or listed in a finite sequence, it is finite.

**Cardinality:**

Calculate the cardinality of the set, which represents the number of elements it contains. If the cardinality is a specific natural number or zero, the set is finite.

**Repetition of Elements:**

Check if the elements of the set repeat in a cyclical manner. If there is a pattern of repetition, the set is likely finite. In contrast, if the elements continue indefinitely without repetition, the set may be infinite.

## Boundlessness:

Consider whether there is an upper limit to the elements of the set. If there is no upper limit and the elements can potentially continue indefinitely, the set is likely infinite.

## Patterns or Structure:

Analyze the structure of the set and look for any patterns or characteristics that suggest finiteness or infiniteness. For example, if the set represents a finite collection of objects or has a well-defined endpoint, it is likely finite.

## Mathematical Analysis:

Utilize mathematical techniques or tools, such as set theory and cardinality concepts, to analyze the set and determine its nature.

## Context:

Consider the context in which the set is defined or used. Sets representing finite collections of real-world objects or bounded mathematical concepts are typically finite, while sets with unbounded or limitless elements may be infinite.

By considering these factors and properties, you can assess whether a set is finite or infinite. However, in some cases, determining the finiteness or infiniteness of a set may require further analysis or mathematical reasoning.

# 1.2. FUNCTIONS

## Definition:

A function f from a set X to a set Y, denoted by $f : X \rightarrow Y$, is a relation from X (the domain) to Y (the co-domain) that satisfies two properties.

i      Every element in X is related to some element in Y.

ii     No elements in X is related to more than one element in Y.

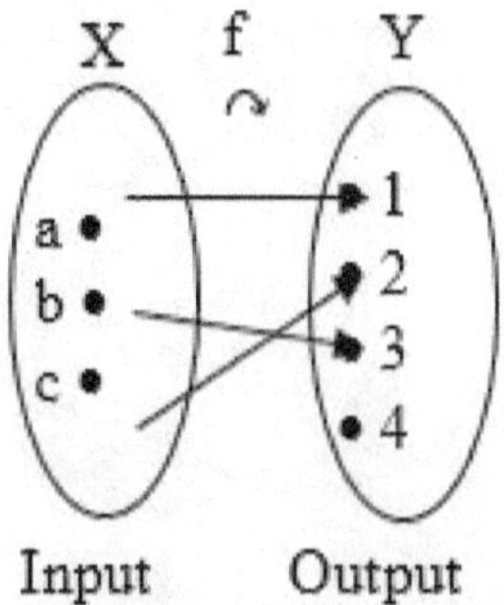

Input          Output

## A Function Relates an Input to an Output

Let say there is some element x in X for which there is a unique element y in Y.

"f sends X to Y" OR "f maps X to Y" OR $f : X \to Y$.

This unique element y is also denoted by f(x) and is called "f of x" OR "Value of f at x" OR the image of x under f.

What is the image of 'a' under f in this arrow diagram ?

Image of 'a' under f is 1

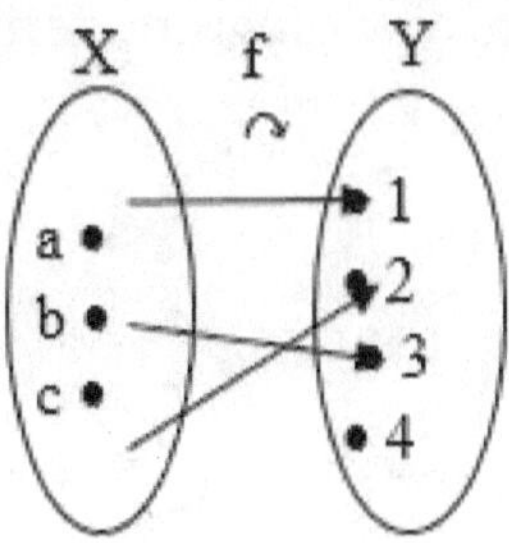

The set of all values of f is called the range of f.

Range of f = {Y ∈ y / y = f(x), for some x in X}

What is the range of f?

Range of f = {1, 2, 3}

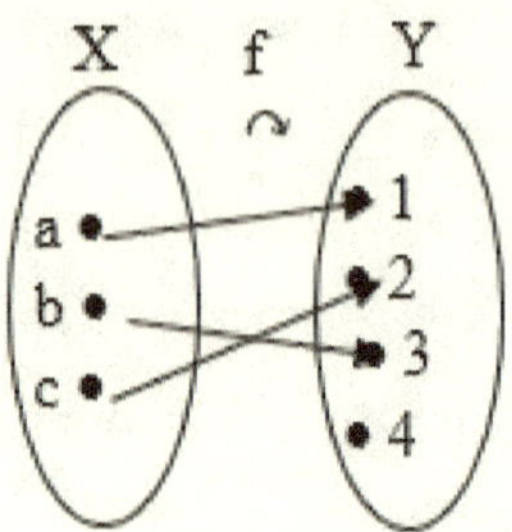


It is also quite obvious that given an element y in Y, there may exist elements in X with Y as their image.

If $f(x) = y$, then x is called a preimage of y OR inverse image of y.

What is the inverse image of 3?

Inverse image of Y = {x ∈ X \ f(x) = y}

Inverse image of 3 = {b}.

**Example 1 :**

Which of the following arrow diagram define a function from X = {a, b, c} to Y = {1, 2, 3, 4} ?

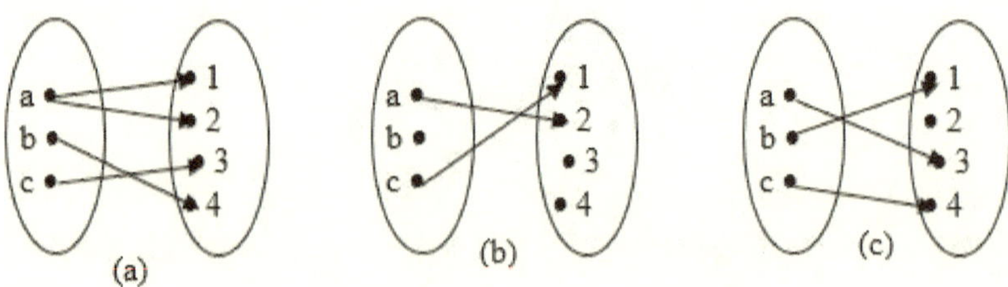

- In (a) there is an element in the domain which is related to more than one element in the codomain. Hence (a) is not a function.

- In (b), not every element of the domain is related to some element of the co-domain. Hence (b) is not a function.

- Only (c) defines a function because both properties of a function are satisfied.

**Example 2 :**

Let X = {1, 2, 5} and Y = {s, t, u, v}. Define f : X → Y by the following arrow diagram.

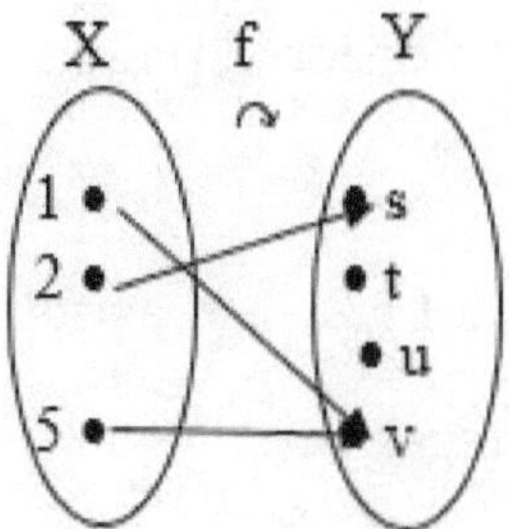


List down the domain of f and the co-domain of f.

Domain of f  = {1, 2, 5}

Co-domain of f = {s, t, u, v}

Find f(1), f(2) and f(5).

f(1) = v

f(2) = s

f(5) = v

What is the range of f ?

Range of f = {s, v}

Is 2 an inverse image of s ?

Yes.

Is 1 an inverse image of u ?

No, there is no arrow pointing to u from 1.

What is the inverse image of s ? of u ? of v ?

Inverse image of s = {2}

Inverse image of u = $\phi$

Inverse image of v = {1, 5}

Represent f as a set of ordered pairs f = {(1, v), (2, s), (5, v)}.

## Example 3 :

Define a unique function f : Z → Z as follows :

For each positive integer n,

$$f(n) = \text{addition of positive divisors of n.}$$

## Find the following :

a)  f(1)          b)  f(15)          c)  f(17)

d)  f(5)          e)  f(18)          f)  f(21)

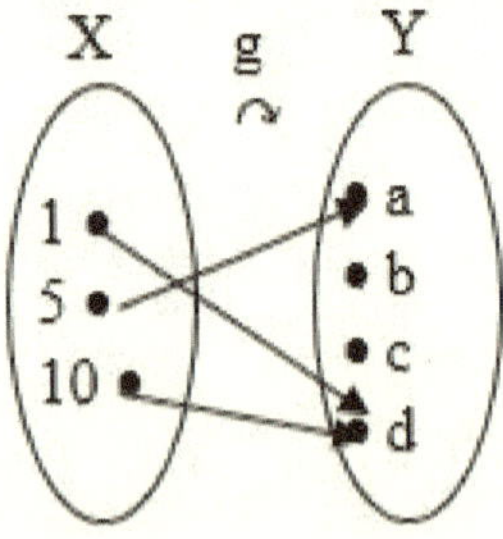

a) $f(1) = 1$

b) $f(15) = 1 + 3 + 5 + 15 = 24$

c) $f(17) = 1 + 17 = 18$

d) $f(5) = 1 + 5 = 6$

e) $f(18) = 1 + 2 + 3 + 6 + 9 + 18 = 39$

f) $f(21) = 1 + 3 + 7 + 21 = 32$

**Try it Yourself:**

Let $X = \{1, 5, 10\}$ and $Y = \{a, b, c, d\}$

Define $g : X \to Y$ by the following arrow diagram.

a)  Write the domain and co-domain of g.

b)  Find $g(1)$, $g(5)$ and $g(10)$.

c)  What is the range of g ?

d)  Is 5 is an inverse image of a ?

e)  Is b an inverse image of 1 ?

f)  What is the inverse image of a and d ?

g)  Represent f as a set of ordered pairs.

## Types of Functions

### 1)  One to One Function

A one to one function is a function in which each element of a range is the image of the at most one element of the domain.

OR

When represented in an arrow diagram, a one-to-one function ensures that each element in the domain corresponds to a unique element in the co-domain.

### Formal definition :

Let f denote a function mapping from set X to set Y. The function f is one-to-one, or injective, if and only if, for all elements $x_1$ and $x_2$ in set X,

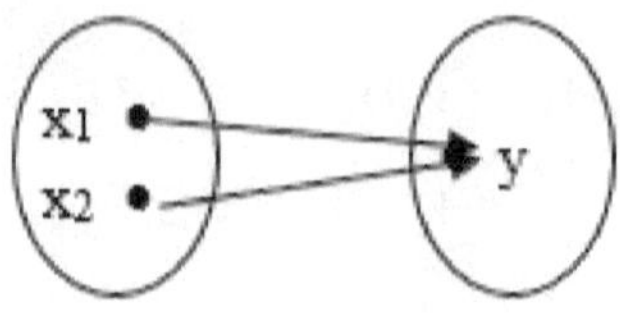

If $f(x_1) = f(x_2)$ then $x_1 = x_2$   OR

If $x_1$ Not equal to $x_2$ then $f(x_1)$ not equal to $f(x_2)$

If $x_1$ and $x_2$ are pointing to the same element y in the co-domain then $x_1 = x_2$.

### Symbolically :

f : X → Y is one to one ⟺ ∀ $x_1$, $x_2$ ∈ X, if $f(x_1)=f(x_2)$ then $x_1 = x_2$

### Problem :

Determine whether each of these functions {a, b, c, d} to itself is one to one.

a)   f(a) = b, f(b) = a, f(c) = c, f(d) = d

b)   f(a) = b, f(b) = b, f(c) = d, f(d) = c

c)   f(a) = d, f(b) = b, f(c) = c, f(d) = d

### Solution :

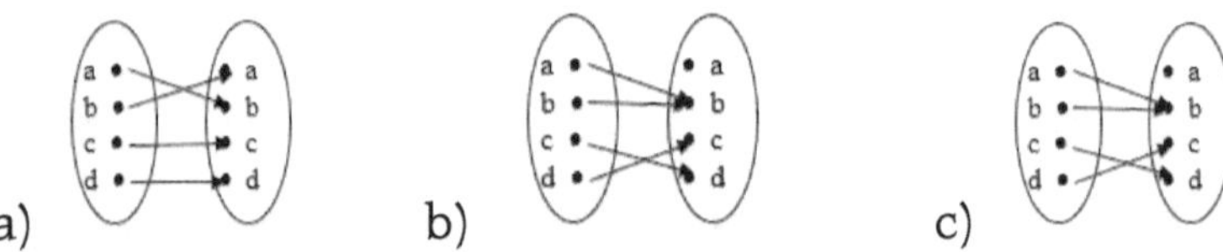

This  a  one  to  one  This is not a one   This is not a one to
function                to one function       one function

## 2)   Onto Function

### Definition :

Let say a function f is defined from a set X to set Y.

This function f is onto (or surjective), if and only if, given any element y in Y, there exist an element in x in X, such that f(x) = y.

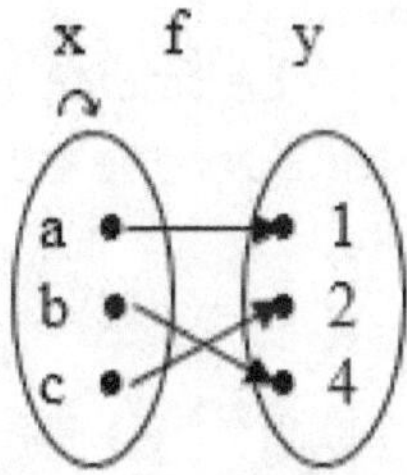

Every element in the co-domain must have an arrow pointing to it from some element of the domain.

### More formally,

$f : X \rightarrow Y$ is onto $\Leftrightarrow \forall y \in Y, \exists x \in X$ such that $f(x) = y$.

y in Y must be the image of some element x in X.

### Example 1 :

Let X = {1, 2, 3, 4} and Y = {a, b, c}. Define H : X → Y as follows :

$$H(1) = c, H(2) = a, H(3) = c, H(4) = b.$$

Define K : X →Y as below :

$$K(1) = c, K(2) = b, K(3) = b \text{ and } K(4) = c$$

is either H or K onto ?

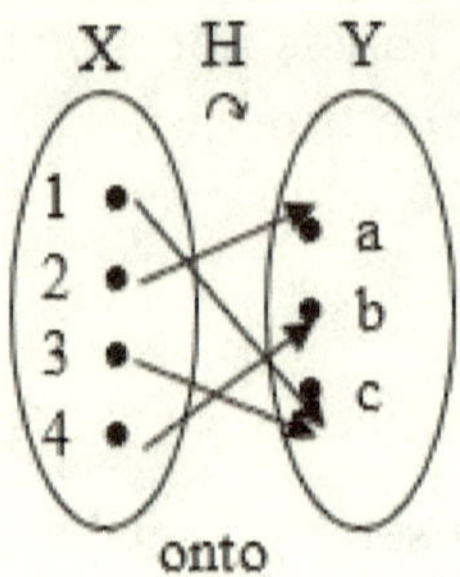

onto

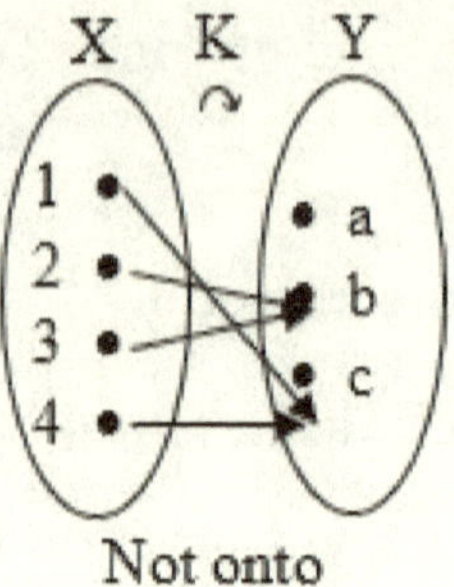

Not onto

## 3)  One to One Correspondence

**Definition :**

The function f is said to be one to one correspondence (or a bijection) if it is both one to one and onto.

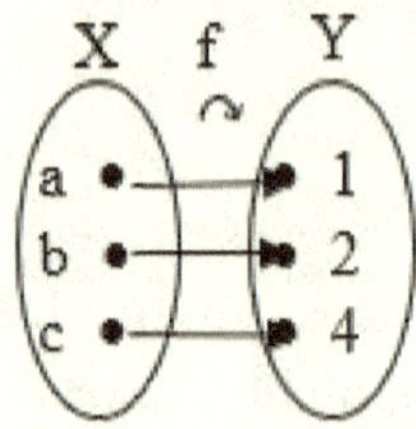

f is both one – to – one and onto

**More Examples**

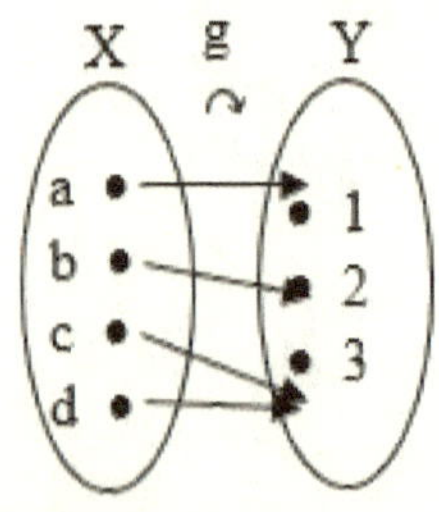

Onto but not one to one

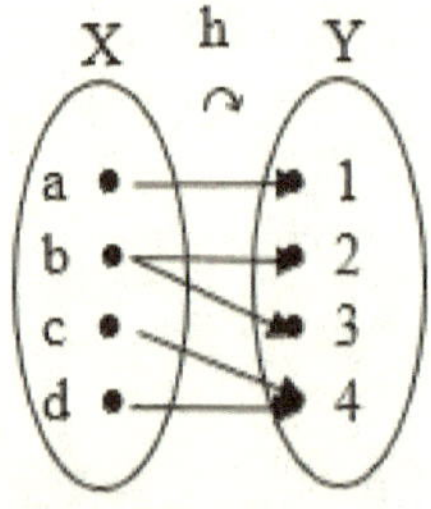

Not a function

**Problem :**

Let A = {a, b} and S = {0 0, 0 1, 1 0, 1 1}. A function f is defined from set P(A) to S where P(A) denotes the power set of A = {φ, {a}, {b}, {a, b}} as follows :

Given any subset X of A.

If element 'a' is in X, then write 1 in the first position of string f(x), if not, then write O in the first position of string f(x).

If element 'b' is in X, then write 1 in the second position of string f(x). If not, then write O in the second position of string f(x).

If  the function 'f' is  a one-to-one correspondence?

| Subset of {a, b} | String in S |
|---|---|
| φ | 0 0 |
| {a} | 1 0 |
| {b} | 0 1 |
| {a, b} | 1 1 |

**Solution :**

A = {a, b} and S = {0 0, 0 1, 1 0, 1 1}

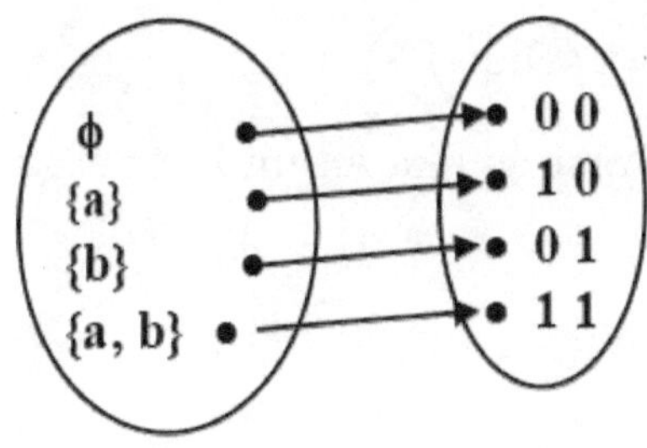


Therefore, f is a one-to-one correspondence.

Let say f is a function from a set X to itself.

This mean f : X → X

If X is finite, then f is one-to-one if and only if f is onto.

**Example :**

Let say f is a function from a set A = {1, 2, 3} to itself.

Let say f is one-to-one.

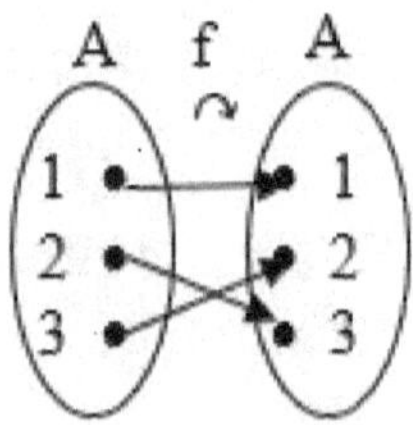

Let say f is onto.

Then definitely, f is one-to-one. Therefore, f is one-to-one if and only if f is onto.

Always true for finite sets. May not be true for infinite sets

4) **Inverse Functions**

**Definition :**

Let f : X → Y is a one-to-one correspondence. Then there is a function $f^{-1}$ : Y → X that is defined as follows :

$f^{-1}(y)$ = unique element x in X such that f(x) = y.

Note : $f^{-1}(y)$ = x when f(x) = y.

Inverse is like an undo operation

## Example 1 :

Consider the following arrow diagram.

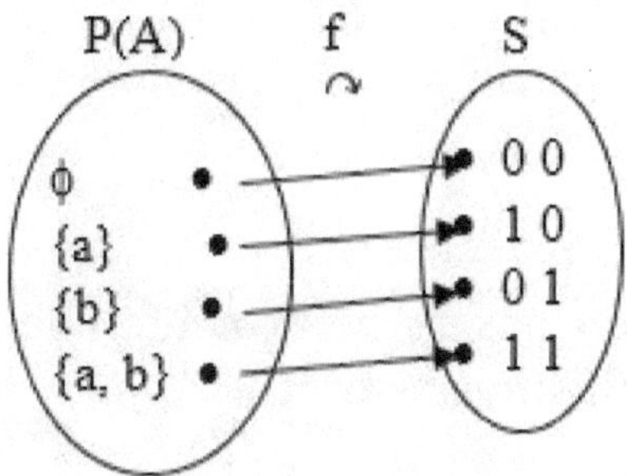

What is the inverse of f ?

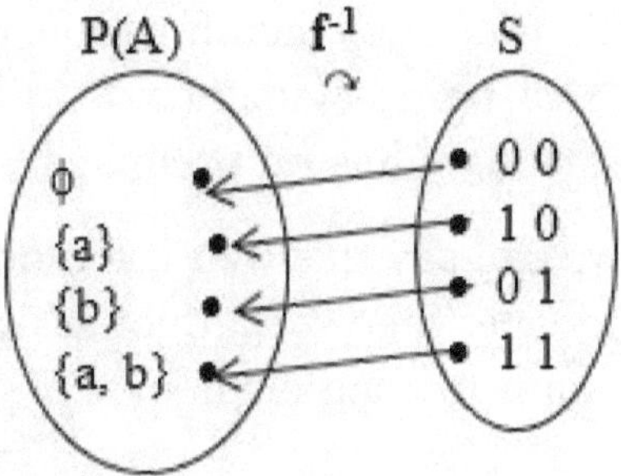

Inverse of f

## Example 2 :

Let $f : R \to R$ is defined as follows :

$$f(x) = 4x - 1$$

The above function is a one-to-one correspondence. Find its inverse function.

**Solution :**

For any y in R,

$$f^{-1}(y) = \text{unique element x in X such that } f(x) = y$$

$$\Rightarrow f(x) = y$$

$$\Rightarrow 4x - 1 = y$$

$$\Rightarrow x = \frac{y+1}{4}$$

$$\Rightarrow f^{-1}(y) = \frac{y+1}{4}$$

Why a function has to be a one-to-one correspondence in order to find the inverse of it ?

If a function f is not one-to-one, then some element y in the co-domain must be the image of more than one element in the domain. This means no unique element.

If a function f is not onto, then for some element y in the co-domain there will be no element x in the domain such that $f(x) = y$. This means y has no preimage.

In summary, we can say this that if a function f is not one-to-one correspondence, then we cannot assign a unique element x from the domain to each element y of the co-domain such that $f(x) = y$.

Due to this reason, a one-to-one correspondence is called invertible.

## 5)   Composition of Functions

Let say we have two functions f and g defined as follows :

$$f(x) = x + 1$$

$$g(x) = x^2$$

Think of f and g as two machines which are accepting some input and producing same output.

Input of machine g is coming from the output of machine of f.

Combining functions in this way is called compositing them and the resulting function is called composition of the two functions.

**Definition :**

Let $f : X \rightarrow Y$ and $g : Y \rightarrow Z$ be function with the property that the range of f is a subset of the domain of g.

The new function $g \circ f : X \rightarrow Z$ is defined as follows :

$$(g \circ f)(x) = g(f(x)) \text{ for all } x \in X$$

The function g o f is called the composition of f and g.

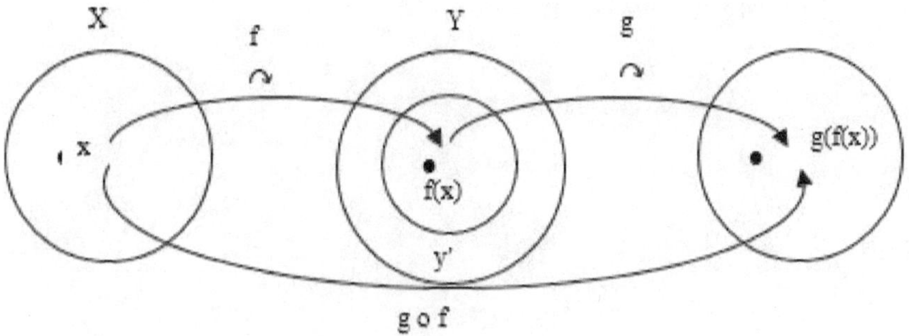


**Commutative property of composition :**

**Example :**

Let $f : Z \rightarrow Z$ and $g : Z \rightarrow Z$ be the two functions where $f(x) = x + 1$ and $g(x) = x^2$.

What is g o f and f o g ?

**Solution :**

$$(g \circ f)(x) = g(f(x)) = g(x + 1) = (x + 1)^2$$

$$(f \circ g)(x) = f(g(x)) = f(x^2) = x^2 + 1$$

Is $g \circ f = f \circ g$ ?

**Solution :**

Two functions are said to be equal, if and only if, they always take the same values.

$$g \text{ o } f\,(2) \;=\; (2 + 1)^2 \;=\; 9$$

$$f \text{ o } g\,(2) \;=\; 2^2 + 1 \;=\; 5$$

$$g \text{ o } f \;\neq\; f \text{ o } g$$

Fact : Composition of functions is not a commutative operation.

Let the functions $f : X \rightarrow Y$ and $g : Y \rightarrow Z$ are defined by the following arrow diagrams :

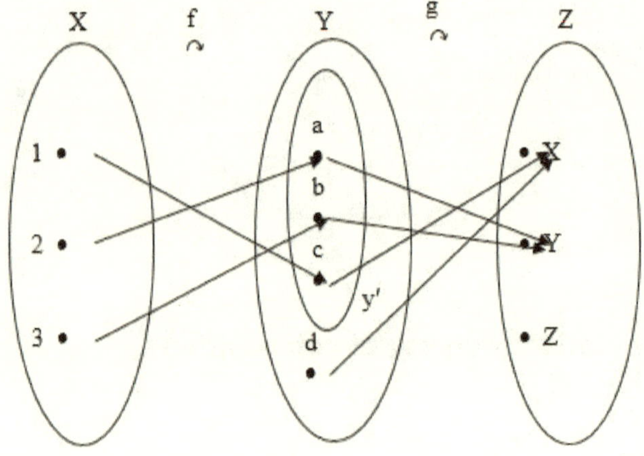


Draw the arrow diagram for g o f.

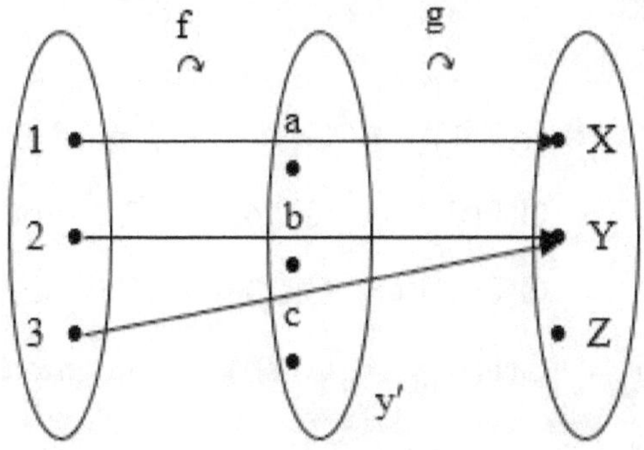

(g o f) (1) = g(f(1))    (g o f) (2) = g(f(2))    (g o f) (3) = g(f(3))

= g(c)                        = g(a)                        = g(b)

=   x                          = Y                          = y

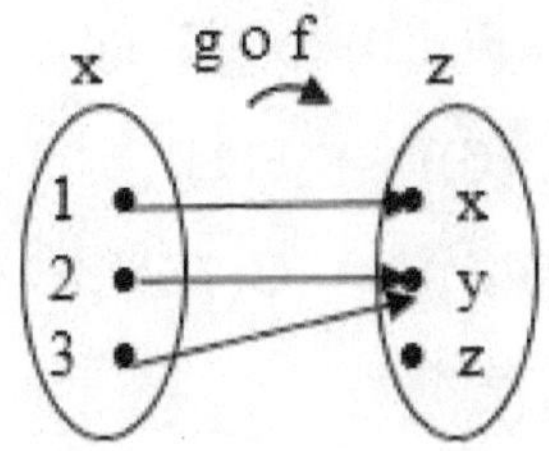

(g o f) (1)   =   g(f(1))   =   g(c)   = x

(g o f) (2)   =   g(f(2))   =   g(a)   = y

(g o f) (3)   =   g(f(3))   =   g(b)   = y

What is the range of g o f ?

Range of g o f is {x, y}.

**Problem 1 :**

Define function $F : Z \to Z$ and $G : Z \to Z$ by the rules F(a) = 7a and G(a) = a mod 5 for all integers a.

Find (G o F) (1), (G o F) (2), (G o F) (3) and (G o F) (4).

**Solution :**

(G o F) (1) ?

(G o F) (1) $=$ G(F(1)) $=$ G(7) $=$ 7 mod 5 $=$ $\boxed{2}$

(G o F) (2) $=$ G(F(2)) $=$ G(14) $=$ 14 mod 5 $=$ $\boxed{4}$

(G o F) (3) $=$ G(F(3)) $=$ G(21) $=$ 21 mod 5 $=$ $\boxed{1}$

(G o F) (4) $=$ G(F(4)) $=$ G(28) $=$ 28 mod 5 $=$ $\boxed{3}$

**Problem 2 :**

Define function F : Z $\rightarrow$ Z and G : Z $\rightarrow$ Z by the rules F(a) $=$ $a^2$ and G(a) $=$ a mod 5 for all integers a.

Find (F o G) (12), (G o F) (12), (F o G) (6) and (G o F) (6).

**Solution :**

(G o F) (12) $=$ G(F(12)) $=$ G(144) $=$ 144 mod 5 $=$ $\boxed{4}$

(F o G) (6) $=$ F(G(6)) $=$ F(1) $=$ $1^2$ $=$ $\boxed{1}$

(G o F) (6) $=$ G(F(6)) $=$ G(36) $=$ 36 and 5 $=$ $\boxed{1}$

Is F o G $=$ G o F ?

**Solution :**

(F o G) (9) $=$ F(G(9)) $=$ F(4) $=$ $\boxed{16}$

(G o F) (9) $=$ G(F(9)) $=$ G(81) $=$ $\boxed{1}$

$$F \text{ o } G \neq G \text{ o } F$$

## 1.3. RELATIONS

**Definition :**

Let A and B be two sets.  A binary relation R from A to B is a subset of A $\times$ B or R $\subseteq$ A $\times$ B.

**Recall :**

$$A \times B = \{(a, b) \mid a \in A \text{ and } b \in B\}.$$

Usually we use the notation a R b to denote $(a, b) \in R$.

a R b is used to denote $(a, b) \notin R$.

**Example :**

Let A = {1, 2, 3} and B = {0, 1, 2, 4}

$$A \times B = \{(1, 0), (1, 1), (1, 2), (1, 4), (2, 0), (2, 1), (2, 2),$$
$$(2, 4), (3, 0), (3, 1), (3, 2), (3, 4)\}$$

Let say R is the relation where $(a, b) \in R$ if and only if a = b then R = {(1, 1), (2, 2)} and

$$R \subseteq A \times B.$$

Graphically representation of ordered pairs.

Arrows are used to represent ordered pairs of relation R

A relation on a set A is a relation from A to A.

**Example :**

Let R = {(a, b) | a divides b}

A = {1, 2, 3, 4, 5, 6}

R = {(1, 1), (1, 2), (1, 3), (1, 4), (1, 5), (1, 6), (2, 2), (2, 4),
(2, 6), (3, 3), (3, 6), (4, 4), (5, 5), (6, 6)}

## Graphical representation of relation R

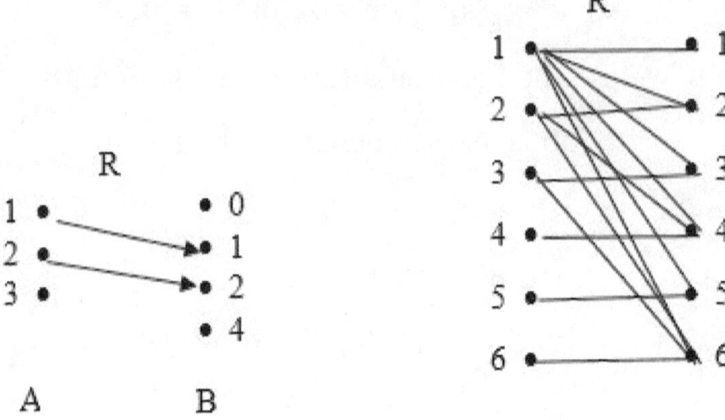

## Number of relations on a set with n elements

- A relation on a set A is a subset of A × A.

- Set A contains n elements and A × A contains $n^2$ elements.

- We are interested in listing down all the subsets of A × A means we are interested in finding the power set of A.

- We know that p(A) is the power set of A and let say A has n elements then p(A) must have $2^n$ elements.

- Therefore, p(A × A) must have $2^{n^2}$ elements.

- Hence, number of relations on a set A with n elements $= |p(A × A)| = 2^{n^2}$

## 1.4. PROPERTIES OF BINARY RELATIONS

1.   **Reflexive Relation :**

A relation R on a set A is called reflexive if $(a, a) \in R$ for every element $a \in A$.

In other words, $\forall a ((a, a) \in R)$.

**Example :**

Let A = {1, 2, 3, 4}

$$R_1 = \{(1, 1), (1, 2), (2, 2), (2, 3), (3, 3), (4, 4)\}$$

Relation $R_1$ is reflexive because it contains all ordered pairs of the form (a, a) for every element a $\in$ A i.e., $R_1$ has (1, 1), (2, 2), (3, 3), (4, 4).

$$R_2 = \{(1, 1), (1, 2), (2, 1), (2, 2), (3, 1), (4, 4)\}$$

Relation $R_2$ is not reflexive because the ordered pair (3, 3) is not in $R_2$.

**2. Irreflexive Relation :**

A relation R on a set A is called irreflexive if $\forall a \in A, (a,a) \notin R$.

**Example :**

A = {1, 2, 3, 4}

$R_3 = \{(1, 2), (2, 1), (3, 3), (4, 4)\}$ is not irreflexive because (3, 3) and (4, 4) is there in $R_3$.

$R_4 = \{(1, 2), (2, 1)\}$ is irreflexive because $\forall$ a $\in$ A. (a, a) $\notin R_4$.

**3. Symmetric Relation :**

A relation R on a set A is called symmetric if (b, a) $\in$ R holds when (a, b) $\in$ R for all  a, b $\in$ A.

In other words, relation R on a set A is symmetric if $\forall a \ \forall b$ ((a, b) $\in$ R $\Rightarrow$ (b, a) $\in$ R).

**Example :**

Relation $R_5 = \{(1, 1), (1, 2), (2, 1), (2, 2)\}$ is symmetric because for every

$$(a, b) \in R_5, (b, a) \in R_5$$

like (1, 2) (2, 1) is in $R_5$.

There is no need to check for $(1, 1)$, $(2, 2)$.

Relation $R_6 = \{(1, 1), (1, 2), (1, 3), (1, 4)\}$ is not symmetric because for $(1, 2)$ there is no $(2, 1)$ in $R_6$.

Same is true for $(1, 3)$ and $(1, 4)$.

4.   **Antisymmetric Relation :**

A relation R on a set A is called antisymmetric if

$$\forall a \; \forall b \; ((a, b) \in R \wedge (b, a) \in R \Rightarrow (a = b)).$$

Whenever we have $(a, b)$ in R, we will never have $(b, a)$ in R until or unless $(a = b)$.

**Example :**

Relation $R_7 = \{(1, 1), (2, 1)\}$ on set A is antisymmetric because $(2, 1)$ is in $R_7$ but $(1, 2)$ is not in $R_7$.

5.   **Transitive Relation :**

A relation R on a set A is called transitive if

$$\forall a \; \forall b \; \forall c \; ((a, b) \in R \wedge (b, c) \in R) \Rightarrow (a, c) \in R).$$

**Example :**

$A = \{1, 2, 3, 4\}$

$R_8 = \{(2, 1), (3, 1), (3, 2), (4, 4)\}$ is transitive because $(3, 2)$, $(2, 1)$ and $(3, 1)$ are there in $R_8$.

$R_9 = \{(2, 1), (1, 3)\}$ is not transitive as $(2, 1)$ and $(1, 3)$ are there in $R_9$ but there is no $(2, 3)$ in relation $R_9$.

6.   **Asymmetric Relation :**

A relation R on a set A is called asymmetric if

$$\forall a \; \forall b \; ((a, b) \in R \Rightarrow (b, a) \notin R).$$

**Example :**

A = {1, 2, 3, 4}

$R_{10}$ = {(1, 1), (1, 2),. (1, 3)} is not an asymmetric relation because of (1, 1).

$R_{11}$ = {(1, 2), (1, 3), (2, 3)} is an asymmetric relation.

**Summary :**

| Relation | Property |
|---|---|
| 1. Reflexive | $\forall a\ ((a, a) \in R)$ |
| 2. Irreflexive | $\forall a\ ((a, a) \notin R)$ |
| 3. Symmetric | $\forall a\ \forall b\ ((a, b) \in R \Rightarrow (b, a) \in R)$ |
| 4. Antisymmetric | $\forall a\ \forall b\ (((a, b) \in R \wedge (b, a) \in R) \Rightarrow (a = b))$ |
| 5. Asymmetric | $\forall a\ \forall b\ ((a, b) \in R \Rightarrow (b, a) \notin R)$ |
| 6. Transitive | $\forall a\ \forall b\ \forall c\ (((a, b) \in R \wedge (b, c) \in R) \Rightarrow (a, c) \in R$ |

**Problem :**

Determine whether relation R on the set of all real numbers is reflexive, symmetric, antisymmetric and / or transitive where (x, y) $\in$ R if and only if

a)   x + y = 0

b)   x – y is a rational number

c)   xy = 0

d)   x = 1 or y = 1

**Solution :**

a)    R = {(x, y) | x + y = O}

1.  **Reflexive :**

    $\forall a \in R ((a, a) \in R)$

    Not reflexive. Only true for (O, O).

2.  **Symmetric :**

    $\forall a \ \forall b \in R ((a, b) \in R \to (b, a) \in R$

    If a + b = O then b + a = O

    Therefore, R is symmetric.

3.  **Antisymmetric :**

    $\forall a \ \forall b \in R (((a, b) \in R \wedge (b, a) \in R) \to (a = b))$

    Not antisymmetric.

    1 + (− 1) = O and − 1 + 1 = O but 1 ≠ − 1

4.  **Transitive :**

    $\forall a \ \forall b \ \forall c (((a, b) \in R \wedge (b, c) \in R) \Rightarrow (a, c) \in R)$

    Not transitive.

    1 + (− 1) = O and (−1) + 1 = O but 1 + 1 ≠ O.

b)    R = {(x, y) | x − y is a rational numbers}

1.  **Reflective :**

    a − a = O is a rational number.

2.  **Symmetric :**

    If a − b is a rational number then b − a = − (a − b)
    is also a rational number.

**3. Antisymmetric :**

Not antisymmetric.

$3 - 2$ and $2 - 3$ are both rational numbers but $3 \neq 2$.

**4. Transitive :**

If $a - b$ is a rational number and $b - c$ is also a rational number, then $a - c = (a - b) + (b - c)$ is also a rational number.

c)   $R = \{(x, y) \mid xy = 0\}$

**1. Reflexive :**

Not reflexive.

$a\,a = a^2 = 0$ when $a = 0$.

**2. Symmetric :**

$a\,b = 0$ then $b\,a = 0$

**3. Antisymmetric :**

Not antisymmetric.

$1 \times 0 = 0$ and $0 \times 1 = 0$ but $1 \neq 0$

**4. Transitive :**

Not transitive.

$-1 \times 0 = 0$ and $0 \times 2 = 0$ but $-1 \times 2 \neq 0$

d)   $R = \{(x, y) \mid x = 1 \text{ or } y = 1\}$

**1. Reflexive :**

Not reflexive.

$(0, 0)$ is not in R.

**2. Symmetric :**

If $(a,b) \in R$ then $a = 1$ or $b = 1$.

This means $(b, a) \in R$.

3. **Antisymmetric :**

Not antisymmetric.

If a = 1 and b = O then (a, b) $\in$ R and (b, a) $\in$ R but a $\neq$ b.

4. **Transitive :**

If $a_1$ = 1 and $b_1$ = O and $a_2$ = O and $b_2$ = 1 where $(a_1, b_1)$, $(a_2, b_2)$ $\in$ R.

Then $(a_1, b_2)$ $\in$ R as $a_1$ = 1 and $b_2$ = 1.

This is transitive.

## 1.5. CLOSURE

1) **Reflexive Closure :**

Let say we have a binary relation R.

R={(1, 1), (2, 2), (2, 3)} defined on a set A = {1, 2, 3}.

The relation is not reflexive.

Smallest reflexive relation that contains R must include the ordered pair (3, 3).

$$R_{New} = \{(1, 2), (2, 2), (3, 3), (2, 3)\}$$

**Definition :**

Reflexive closure of a binary relation R on a set A is the smallest reflexive relation of the set A that contains R.

Reflexive closure of R is usually denoted by $R_r^+$

$$R_r^+ = R \cup \{(a, a) \mid a \in A\}$$

**Problem :**

Let R be the relation on the set {0, 1, 2, 3} containing the ordered pairs (0, 1), (1, 1), (1, 2), (2, 0), (2, 2) and (3, 0). Find the reflexive closure of R.

**Solution :**

R = {(0, 1), (1, 1), (1, 2), (2, 0), (2, 2), (3, 0)}

A  =  {0, 1, 2, 3}

Reflexive closure of R

$$R_r^+ \ = \ R \cup \{(a,\, a) \mid a \in A\}$$

$R_r^+ = \{(0, 1), (1, 1), (1, 2), (2, 0), (2, 2), (3, 0), (0, 0), (3, 3)\}$

The relation 1) contains R ; 2) is reflexive ;  3)  is minimal.

**2)   Symmetric Closure :**

Symmetric closure of a binary relation R on a set A is the smallest symmetric relation on a set A that contains R.

$$R_s^+ \ = \ R \cup \{(b,\, a) \mid (a,\, b) \in R\}$$

**Problem :**

Let R be the relation on the set {0,1, 2, 3} containing the ordered pairs (0, 1), (1, 1), (1, 2), (2, 0), (2, 2) and (3, 0). Find the symmetric closure of R.

**Solution :**

R = {(0, 1), (1, 1), (1, 2), (2, 0), (2, 2), (3, 0)}

A = {0, 1, 2, 3}

$R_s^+ = R \cup \{(b,\, a) \mid (a,\, b) \in R\}$

$R_s^+ = \{(0, 1), (1, 1), (1, 2), (2, 0), (2, 2), (3, 0), (1, 0),$
$(2, 1), (0, 2), (0, 3)\}$

**3)  Transitive Closure :**

Transitive closure of a binary relation R on a set A is the smallest transitive relation on a set A that contains R.

$$R_t^+ \;=\; R \cup \{(a, c) \mid (a, b) \in R \wedge (b, c) \in R\}$$

**Problem :**

Let R be the relation on a set {1, 2, 3} containing the ordered pairs (1, 1), (2, 3) and (3, 1). Find the transitive closure of R.

**Solution :**

R = {(1, 1), (2, 3), (3, 1)}

A = {1, 2, 3}

$$R_t^+ \;=\; R \cup \{(a, c) \mid (a, b) \in R \wedge (b, c) \in R\}$$

$$R_t^+ \;=\; \{(1, 1), (2, 3), (3, 1), (2, 1)\}$$

**Solved Problems**

**Problem 1 :**

Let R be the relation {(a, b) | a ≠ b} on the set of integers. What is the reflexive closure of R.

**Solution :**

R = {(a, b) | a ≠ b}

Relation R is defined on set of integers.

$$R_r^+ \;=\; R \cup \{(a, a) \mid a \in A\} \quad [\text{A is the set of all integers}]$$

$$\;=\; \{(a, b) \mid a \neq b\} \cup \{(a, b) \mid a = b\}$$

This means all pairs of integers must be included in the reflexive closure of

$$R \;=\; \{(a, b) \mid a, b \in Z\} \;=\; Z \times Z$$

Therefore, $R_r^+ = Z \times Z$

**Problem 2 :**

Let R be the relation $\{(a, b) \mid a \text{ divides } b\}$ on the set of integers.

What is the symmetric closure of R ?

**Solution :**

$R = \{(a, b) \mid a \text{ divides } b\}$

$A = \text{Set of integers} = Z.$

$$R_s^+ = R \cup \{(b, a) \mid (a, b) \in R\}$$
$$= \{(a, b) \mid a \text{ divides } b\} \cup \{(b, a) \mid a \text{ divides } b\}$$
$$= \{(a, b) \mid a \text{ divides } b\} \cup \{(a, b) \mid b \text{ divides } a\}$$
$$= \{(a, b) \mid a \text{ divides } b \text{ or } b \text{ divides } a\}$$

## 1.6. EQUIVALENCE RELATIONS

**Definition :**

A relation R on a set A is an equivalence relation iff R is reflexive, symmetric and transitive.

**Example :**

$A = \{0, 1, 2, 3\}$

$R_1 = \{(0, 0), (1, 1), (2, 3), (3, 3)\}$

Is $R_1$ an equivalence relation ? Yes.

$R_2 = \{(0, 0), (0, 2), (2, 0), (2, 2), (2, 3), (3, 2), (3, 3)\}$

Is $R_2$ an equivalence relation ?

Is $R_2$ reflexive ?

No. Because $(1, 1)$ is not a member of $R_2$.

$R_3 = \{(0, 0), (1, 1), (1, 2), (2, 1), (2, 2), (3, 3)\}$

Is $R_3$ an equivalence relation ?

Ask yourself :     Is $R_3$ reflexive ?  Yes.

Is $R_3$ symmetric ?  Yes.

Is $R_3$ transitive ?  Yes.

$R_4 = \{(0,0), (0,1), (0,2), (1,0), (1,1), (1,2), (2,0), (2,2), (3,3)\}$

Is $R_4$ an equivalence relation ?

Ask yourself :     Is $R_4$ reflexive ?  Yes.

Is $R_4$ symmetric ?  No.

Therefore $R_4$ is not an equivalence relation.

$R_5 = \phi$  is $R_5$ an equivalence relation ?

Ask yourself :     Is $R_5$ reflexive ? No.

Is $R_5$ symmetric ?

Therefore, $R_5$ is not an equivalence relation.

$R_6 = A \times A$          Is $R_6$ an equivalence relation ?

Ask yourself :     Is $R_6$ reflexive ? Yes.

Is $R_6$ symmetric ? Yes.

Is $R_6$ transitive ? Yes.

Therefore, $R_6$ is an equivalence relation.

## 1.7. PARTIAL ORDERING RELATIONS

Many a time, we use relations to order some or all the elements of a set.

Let say we have a set of all words in the dictionary

$$R = \{(a, b) \mid a \text{ comes before } b\}.$$

Partition comes before relation

(Partition, Relation)

1 2 3 4 5

6 7 8 9 10

Let say we have a set of all integers

$$R = \{(a, b) \mid a < b\}$$

In the previous examples, we have seen how a relation indicates that a certain element precedes the other in the ordering.

These relations are used to order some or all elements of a particular set.

Set of all words in a dictionary   Set of all integers

$R_1 = \{(a, b) \mid a \text{ comes before b}\}$        $R_2 = \{(a, b) \mid a < b\}$

Let say we also add pairs of the form $(a, a)$ in $R_1$ and $R_2$.

$R_1 = \{(a, b) \mid a \text{ comes before b or a is equal to b}\}$  $R_2 = \{(a,b) \mid a \leq b\}$

We have obtained relations $R_1$ and $R_2$ that are reflexive, antisymmetric and transitive.

**Definition of Partial Ordering :**

A relation R on set S is called a partial ordering or partial order if it is

1. Reflexive

2. Antisymmetric

3. Transitive

**Reflexive –**        Each element must be related to itself.

**Antisymmetric –** No. Two elements precede each other (ordering).

**Transitive –** First element related to second and second element related to third implies first and third are related.

A set S together with relation R is called a partially ordered set or POSET.

Denoted by (S, R) or (S, $\leq$)

Where a $\leq$ b means a is related to b.

**Example 1 :**

Show that the relation R = {(a, b) | a $\subseteq$ b} defined on the power set of set S = {1, 2, 3} is a partial order relation.

**Solution :**

Relation R is said to be a partial ordering iff R is :

(i)   Reflexive

(ii)  Antisymmetric

(iii) Transitive

S $=$ {1, 2, 3}

P(S) $=$ {$\phi$, {1}, {2}, {3}, {1, 2}, {2, 3}, {1, 3}, {1, 2, 3}}

R $=$ {(a, b) | a $\subseteq$ b}

**(i)    Reflexivity :** a $\subseteq$ a.

**(ii)   Antisymmetry :** a $\subseteq$ b and b $\subseteq$ a implies a = b.

**(iii) Transitivity :** a $\subseteq$ b and b $\subseteq$ c implies a $\subseteq$ c.

Therefore, R is a partial order relation.

**Example 2 :**

Show that the relation R = {(a, b) | a divides b} defined on the set S = {1, 2, 3, 4, 6} is a partial order relation.

**Solution :**

Relation R is said to be a partial ordering iff R is

    (i)   Reflexive

    (ii)  Antisymmetric

    (iii) Transitive

R = {(a, b) | a divides b}      S = {1, 2, 3, 4, 6}

**(i)**  **Reflexivity :**  a divides a.

**(ii)**  **Antisymmetry :**  a divides b and b divides a.

**(iii)**  **Transitivity :**    implies a = b

                     a divides b and b divided c

                     implies a divides c

But what do we mean by partial in partial ordering ?

Meaning of partial in partial ordering.

The word "partial" in "partial ordering" indicates that not every pair of element in a set is comparable.

Two terms which are important for us to understand :

    (1)   Comparable.

    (2)   Incomparable.

**Difference Between Comparable and Incomparable :**

**Comparable :**

The element a and b of poset (S, R) are called comparable if either a R b or b R a.

**Incomparable :**

The element a and b of poset (S, R) are called incomparable if neither a R b nor b R a.

Remember, we can use (S, ≤) to represent an arbitrary poset where a ≤ b is used to denote

$$(a, b) \in R.$$

**Example :**

In the poset (P(S). ⊆) where S = {1, 2, 3}

$$\{1, 3\} \nsubseteq \{2\} \text{ or } \{2\} \nsubseteq \{1, 3\}$$

Therefore {1, 3} and {2} are incomparable.

On the other hand, {1} ⊆ {1, 3} are comparable.

## 1.8. COUNTING

**1) Product Rule :**

Product rule is applied when a procedure is made up of more than one independent tasks.

**Examples :**

→ How many different licence plates are possible if each plate contains a sequence of three letters followed by three digits ?

→ How many different bit strings are there of length of eight?

→   An office building contains 278 floors and has 37 offices on each floor. How many offices are there in the building?

**Definition :**

If there are m ways to do a specific task and for each of these ways of doing the first task, there are n ways to do the second task, then there are m x n ways to do the procedure.

**Example 1 :**

How many three digit members are possible if the allowable digits are 1, 2 and 3 ?

<u>3</u>     <u>3</u>     <u>3</u>

| 111 | 121 | 131 | 211 | 221 | 231 | 311 | 321 | 331 |
| 112 | 122 | 132 | 212 | 222 | 232 | 312 | 322 | 332 |
| 113 | 123 | 133 | 213 | 223 | 233 | 313 | 323 | 333 |

3 x 3 x 3 = 27

**Example 2 :**

How many different licence plats are possible if each plate contains a sequence of three letters followed by three digits?

<u>26</u>   <u>26</u>   <u>26</u>   <u>10</u>   <u>10</u>   <u>10</u>

26 x 26 x 26 x 10 x 10 x 10

=   17,576,000

**Example 3 :**

How many different bit strings are there of length of eight ?

| 2 | 2 | 2 | 2 | 2 | 2 | 2 | 2 |
|---|---|---|---|---|---|---|---|
| 0 or 1 | 0 or 1 | 0 or 1 | 0 or 1 | 0 or 1 | 0 or 1 | 0 or 1 | 0 or 1 |

$$2 \times 2 \times 2 \times 2 \times 2 \times 2 \times 2 \times 2 = 2^8 = 256$$

**Example 4 :**

An office building contains 27 floors and has 37 offices on each floor. How many offices are there in the building ?

$$27 \times 37 = 999$$

**2)  Sum Rule:**

**Definition  :**

If a task can be done in either one of m ways or one of n ways and none of the set of m ways is same as the set of n ways, then there are m + n ways to do the task.

A student can choose a project from one of three lists. The three lists contain 23, 15 and 19 possible projects respectively. No project is on more than one list. How many possible projects are there to choose from ?

There are three lists of 23, 15 and 19 possible projects and student has to choose one project. There is no overlap too.

$$23 + 15 + 19 = 57 \text{ ways.}$$

How many licence plates can be made using either two letters followed by four digits or two digits followed by four letters?

**Case 1 :**   26    26    10    10    10    10

$$26 \times 26 \times 10 \times 10 \times 10 \times 10 = 6760000$$

**Case 2 :**   10    10    26    26    26    26

$$10 \times 10 \times 26 \times 26 \times 26 \times 26 \ = \ 45697600$$

Total   :     $6760000 + 45697600 \ = \ 52457600$

## 1.9. PIGEONHOLE PRINCIPLE

Suppose there are four pigeons and three pigeonholes and these pigeons want to settle into these pigeonholes.

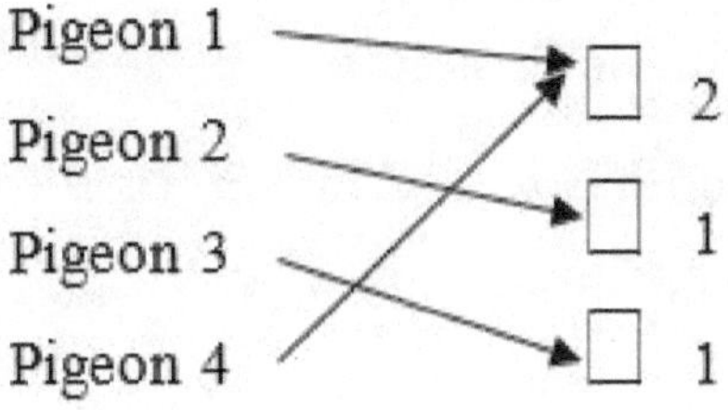


**Theorem  :**

If k is a positive integer and k + 1 or more objects are placed into k boxes, then there is at least one box containing two or more objects. This is the pigeonhole principle.

**Applications :**

Is function f from a set with k + 1 elements to a set with k elements, one to one ?

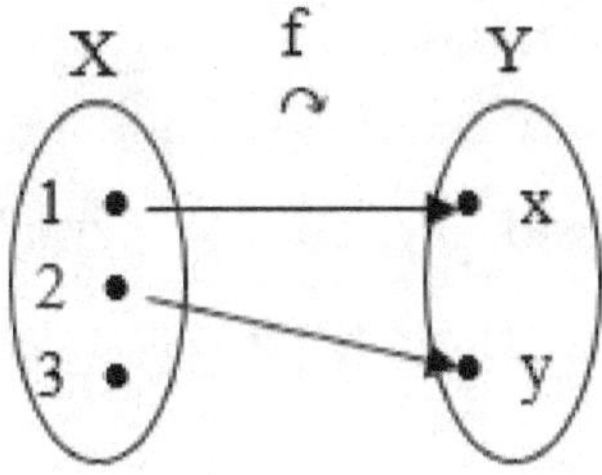


According to the pigeonhole principle, if there are k + 1 objects and the task to place those objects in k boxes, then there must be at least one box with two objects.

How many students must be in a class to guarantee that at least two students receive the same score on the final exam, if the exam is graded on a scale from 0 to 100 points ?

There must be at least 102 students because there are 101 points to distribute and according to pigeonhole principle, if there are only k boxes and k + 1 or more objects has to be placed in them, then at least one box will have two or more objects.

**Generalized Pigeonhole Principle :**

Given m objects and n boxes, if m > n, then there is at least one box with $\left\lceil \frac{m}{n} \right\rceil$ objects.

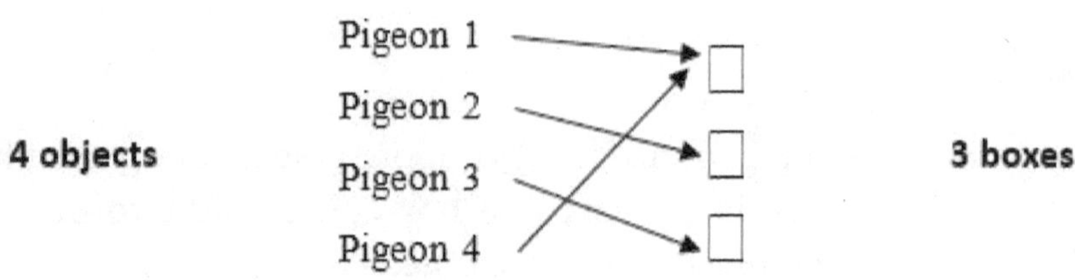


**Problem 1 :**

Among 100 people, how many of them are guaranteed to born in the same month ?

**Solution :**

According to generalized pigeonhole principle, given m objects and n boxes, if m > n, then there is at least one box with $\left\lceil \frac{m}{n} \right\rceil$ objects.

We can treat 100 people as objects and 12 months in a year as boxes.

$$\left\lceil \frac{100}{12} \right\rceil = 9$$

## Problem 2 :

How many cards must be selected from a standard deck of 52 cards to guarantee that at least three cards of the same suit are chosen ?

## Solution :

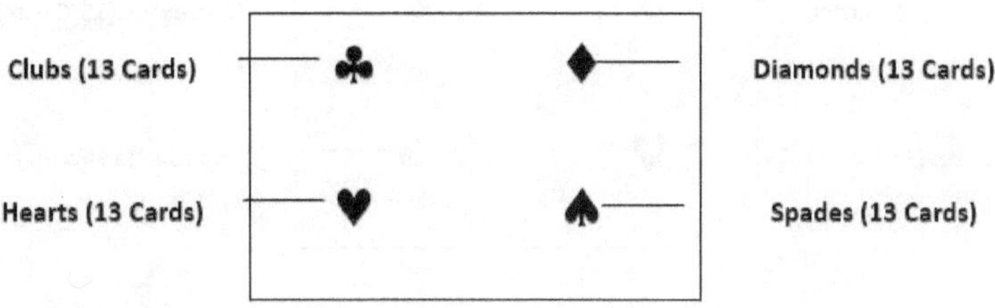


4 suits → clubs, diamonds, hearts and spades.

Let say that each suit is a box and as we select the cards, they will be placed in their corresponding box.

Let say N cards (objects) are selected from the deck of 52 cards.

According to generalized pigeonhole principle : Given m objects and n boxes, if m > n, then there is at least one box with $\left\lceil \frac{m}{n} \right\rceil$ objects.

This means that there is at least one box with $\left\lceil \frac{N}{4} \right\rceil$ cards.

We want $\left\lceil \frac{N}{4} \right\rceil \geq 3$

Smallest value of N = 9

So, minimum 9 cards need to be selected from the deck to guarantee that at least three cards belong to the same suit.

**Problem 3 :**

How many cards must be selected from a standard deck of 52 cards to guarantee that at least three hearts are selected.

**Solution :**

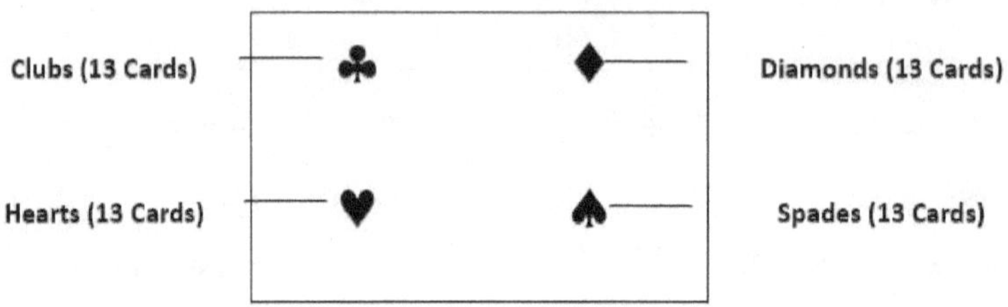


**Worst Case :**

At first, 13 diamonds, 13 clubs and 13 spades are selected. This means first 39 cards are all diamonds, clubs and spades.

Now, next 3 are generalized to be hearts. Therefore, to guarantee that we will get at least three hearts. We need to select total 42 cards from the deck.

## 1.10. PERMUTATION AND COMBINATION

**1)   Permutation**

**Problem 1 :**

A license plate begins with three letters. If the possible letters are A, B, C, D and E, how many different license plates can be formed if no letter is used more than once ?

**Solution :**

$$\underline{5} \qquad \underline{4} \qquad \underline{3}$$

By product rule, 5 x 4 x 3 = 60 license plates.

**Problem 2 :**

Find the number of words, with or without meaning, that can be formed with the letters of the word 'CHAIR' ? (Repetitions are not allowed).

**Solution :**

$$\underline{5} \qquad \underline{4} \qquad \underline{3} \qquad \underline{2} \qquad \underline{1}$$

By product rule, 5 x 4 x 3 x 2 x 1 = 120 words.

**What is a permutation ?**

A permutation of a set of distinct objects is an ordered arrangement of these objects.

Sometimes, we are only interested in ordered arrangements of some objects of the set and not all objects. An ordered arrangements of r objects of a set is called r-permutations.

**Problem 1 :**

A license plate begins with three letters. If the possible letters are A, B, C, D and E. How many different license plates can be formed if no letter is used more than once ?

Here the set consists of letters A, B, C, D and E but we are only interested in ordered arrangements of 3 letters. Hence 3 permutations is 60 license plates.

**Problem 2 :**

Find the number of words, with or without meaning, that can be formed with the letters of the word 'CHAIR' ? (Repetitions are not allowed).

Here we are interested in arranging all letters of the set = {'C', 'H', 'A', 'I', 'R'}.

**Different notations can be used to denote r-permutation :**

The number of r-permutations of a set with n objects is denoted by

$$* \quad p(n, r)$$

$$* \quad n_{P_r}$$

$$* \quad n^P r$$

**Problem 1 :**

A license plate begins with three letters. If the possible letters are A, B, C, D and E, how many different license plates can be formed if no letter is used more than once ?

**Solution :**

$$p(n, r) \ = \ p(5, 3) \ = \ 60$$

$$p(n, n) \ = \ p(5, 5) \ = \ 5! \ = \ 120$$

**Problem 1 :**

A license plate begins with three letters. If the possible letters are A, B, C, D and E, how many different license plates can be formed if no letter is used more than once ?

$$\underline{5} \qquad \underline{4} \qquad \underline{3}$$

By product rule, 5 x 4 x 3 = 60 license plates p(n, r) = p(5, 3) = 60.

**Theorem :**

If n is a positive integer and r is an integer with $1 \leq r \leq n$, then there are

$$p(n, r) \ = \ n(n - 1) (n - 2) - (n - r + 1).$$

**Proof:**

**Problem 1:**

$p(n, r) = p(5, 3) = 5 \times 4 \times 3 = 60$

$p(n, r) = n(n - 1)(n - 2) \ldots (n - (r - 1))$

**Assumption :**

$p(n, 0) = 1.$

**Problem 2 :**

A lock will open if the right choice of 3 numbers (from 1 to 16) is selected. How many lock permutations can be made assuming no number is repeated ?

**Solution :**

$p(n, r) = p(16, 3) = 16 \times 15 \times 14 = 3584$

But how to formulae $p(n, r)$?

By using factorial.

$p(n, n) = p(16, 16) = 16! = 20922789888000$

We want only $16 \times 15 \times 14$ from $16!$

$$\text{Divide by } 13! \quad \frac{16 \times 15 \times 14 \times 13 \times 12 \times \ldots \times 2 \times 1}{13 \times 12 \times 11 \times \ldots \times 1}$$

$$= 3584$$

$$p(16, 3) = \frac{16 \times 15 \times 14 \times 13 \times 12 \times \ldots \times 2 \times 1}{13 \times 12 \times 11 \times \ldots \times 1}$$

In general,     $0 \leq r \leq n$

$$p(n, r) = \frac{n!}{(n-r)!}$$

For $r = 0$

$$p(n, 0) = \frac{n!}{(n-o)!} = \frac{n!}{n!} = 1$$

**Problem 1 :**

List all the permutations of {a, b, c}.

**Solution :**

| | | |
|---|---|---|
| abc | bac | cab |
| acb | bca | cba |

Total 6 permutations.

**Problem 2 :**

In how many different orders can five runners finish a race if no ties are allowed ?

**Solution :**

$$\underline{5} \qquad \underline{4} \qquad \underline{3} \qquad \underline{2} \qquad \underline{1}$$

p(n, n) = p(5, 5) = 5! = 120

**Problem 3 :**

What is the number of possible words that can be made using the word "EASYQUIZ" such that the vowels always come together ?

$$\text{Total}: \ 5\,!\,x\,4\,! \ = \ 2880$$

The vowels are EAUI

These vowels can be arranged in 4 ! ways.

We can now assume the set of vowels 'EAUI' as one character.

Now, we need to arrange 5 characters 'EAUI', 'S', 'Y', 'Q', 'Z'.

$$\underline{5} \qquad \underline{4} \qquad \underline{3} \qquad \underline{2} \qquad \underline{1}$$

All 5 characters can be arranged in 5 ! ways and vowels can be arranged in 4 ! ways

**Solved Problems:**

**Problem 1 :**

What is the number of possible words that can be made using the word 'QUIZ'. Such that the vowels never come together ?

**Solution :**

The total number of words that can be formed from QUIZ are 4 ! = 24 words.

The number of possible words where vowels never come together

= Total words - Number of words where vowels come together

The vowels are UI.

These vowels can be arrange in 2 ! ways.

We can now assume the set of vowels 'UI' as one character.

Now, we  need to arrange 3 characters 'UI', 'Q', 'Z'

$$\underline{3} \qquad \underline{2} \qquad \underline{1}$$

All 3 characters can be arranged in 3 ! ways and vowels can be arranged in 2 ! ways.

$$\text{Total} = 3! \times 2! = 12 \text{ words.}$$

The total number of words that can be formed from QUIZ are 4 ! = 24 words.

The number of possible words where vowels never come together =

Total words – Number of words where vowels come together

$$24 - 12 = 12 \text{ words}$$

**Problem 2 :**

How many 4 digit number can be formed from the digits 1, 2, 3, 4, 5, 6 and 7 which are divisible by 5 when none of the digits are repeated ?

**Solution :**

A number can be divisible by 5 if it ends with either 0 or 5.

As 0 is not available in the given set of digits.

Therefore, a 4 digit number must end with 5.

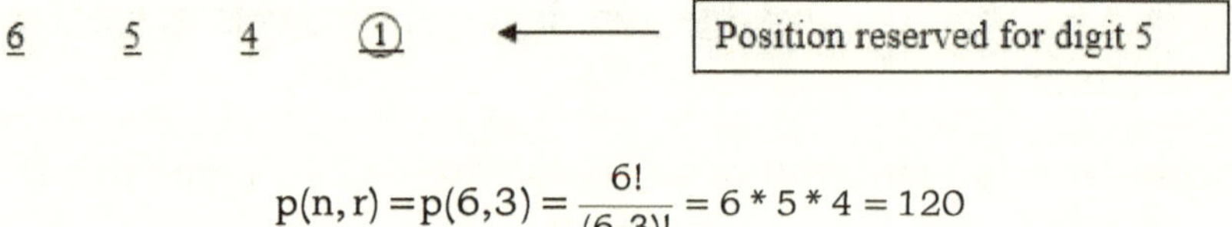

$$p(n,r) = p(6,3) = \frac{6!}{(6-3)!} = 6*5*4 = 120$$

**Problem 3 :**

In how many ways can the alphabets of the word 'DERAIL' be arranged so that the vowels come at the odd positions only ?

**Solution :**

There are 3 vowels 'EAI' and 3 consonants 'DRL'.

| V | C | V | C | V | C |
|---|---|---|---|---|---|
| 3 | 3 | 2 | 2 | 1 | 1 |

$= 3 * 3 * 2 * 2 * 1 * 1$

Total $=$ 36 ways.

## 2) Combination

Combination = Selection

Permutation = Ordered combination

Permutations = Selecting and ordering

**Example :**

Consider a set A with 4 numbers. A={1,2,3,4}. Find p(4, 3).

**Solution :**

Consider a selection of 3 elements 1, 2, 3.

There are 3 ! ways in which elements (1, 2 and 3) can be arranged. r !.

| | | |
|---|---|---|
| 123 | 213 | 312 |
| 132 | 231 | 321 |

The selected elements are 123, 124, 154 and 234.

All the elements need to be arranged in the same way.

Permutations = 4 x 3 ! = 4 x 3 x 2 x 1 = 24 ways.

permutations = Selection x Ordering

$$\text{Selection} = \frac{\text{permutations}}{\text{Ordering}}$$

$$\text{Selection} = \frac{n!(n-r)!}{r!}$$

$$\text{Selection or combination} = \frac{n!}{r!(n-r)!}$$

**Theorem :**

The number of r-combinations of a set with n elements, where n is a nonnegative integer and r is an integer with $0 \le r \le n$, equals

$$c(n,r) = \frac{n!}{r!(n-r)!}$$

## Different notations can be used to denote r-combination :

The number of r-combinations of a set with n objects is denoted by

* $c(n,r)$

* $n_{c_r}$ $\quad\quad \binom{n}{r}$ is called binomial coefficient

* $n c_r$

## Corollary :

If n and r are nonnegative integers with $n \leq r$, then

$c(n, r) \quad = \quad c(n, n-r)$

## Proof:

$$c(n,r) = \frac{n!}{r!(n-r)!}$$

$$c(n,n-r) = \frac{n!}{(n-r)!(n-(n-r))!}$$

$$= \frac{n!}{(n-r)!\,r!}$$

## Solved Problems:

## Problem 1 :

How many ways are there to select 47 cards from a standard deck of 52 cards ?

## Solution :

Apply the corollary,

$c(n, r) \quad = \quad c(n, n-r)$

$c(52, 47) \quad\quad = \quad\quad c(52, 5)$

$$c(52, 5) = \frac{52!}{5!\ 47!} = 2598960$$

## Problem 2 :

How many ways are there to select 3 males and 2 females out of 7 males and 5 females?

## Solution :

The number of ways to select 3 males out of 7 males.

$$c(7, 3) = \frac{7!}{3!\ 4!}$$
$$= \frac{7 \times 6 \times 5}{3 \times 2} = 35$$

The number of ways to select 2 females out of 5 females.

$$c(5, 2) = \frac{5!}{2!\ 3!}$$
$$= \frac{5 \times 4}{2} = 10$$

By the rule of product, 35 x 10 = 350.

## Solved Problems:

## Problem 1 :

In how many ways can a set of two positive integers less than 100 be chosen.

## Solution :

Two positive integers less than 100 ranges from 1 to 99.

$$c(99, 2) = 99C_2 = \frac{99!}{2!\ 97!} = 4851$$

**Problem 2 :**

A coin is flipped eight times where each flip comes up either heads or tails. How many possible outcomes.

a)   are there in total ?

b)   contain exactly three heads ?

c)   contain atleast three heads ?

d)   contain the same number of heads and tails ?

**Solution :**

a)   2 possible outcomes in each flip (either heads or tails).

$$\underline{2} \quad \underline{2} \quad \underline{2} \quad \underline{2} \quad \underline{2} \quad \underline{2} \quad \underline{2} \quad \underline{2}$$

$$\text{H/T} \quad \text{H/T} \quad \text{H/T} \quad \text{H/T} \quad \text{H/T} \quad \text{H/T} \quad \text{H/T} \quad \text{H/T}$$

$$= \quad 2^8 = 256$$

b)   There are total 8 coin flips. Out of these exactly 3 heads should come and 5 tails should appear.

$$8c_3 = \frac{8!}{3! \; 5!} = 56$$

HHHTTTTT

HTHTHTTT

$\cdot$

$\cdot$

$\cdot$

2 possible outcomes in each flip.

c)   $8c_3 + 8c_4 + 8c_5 + ... + 8c_8$

(or)

$$\text{Total} = (8c_0 + 8c_1 + 8c_2)$$

$$= 256 - (1 + 8 + 28)$$

$$= 219$$

d)   This means 4 heads and 4 tails

$$8c_4 = \frac{8\,!}{4\,!\ \ 4\,!}$$

$$= \frac{8 \times 7 \times 6 \times 5 \times 4!}{4 \times 3 \times 2 \times 4!}$$

$$= 70$$

## 1.11. MATHEMATICAL INDUCTION

It is a method of proving a proposition or statement by inductive approach.

**Steps :**

1.  We prove that p(n) is true for some n = $n_0$ generally $n_0$ = 1.

2.  We assume that p(n) is true for n = k.

3.  We have to prove that p(n) is true for n = k + 1.

Then we say p(n) is true for all n $\geq n_0$.

**Problem 1 :**

Show by mathematical induction

$$1^2 + 2^2 + 3^2 + \ldots + n^2 = \frac{n(n+1)(2n+1)}{6}, \text{for } \forall\, n \geq$$

$$\Rightarrow \text{Let } p(n) : 1^2 + 2^2 + 3^2 + \ldots + n^2 = \frac{n(n+1)(2n+1)}{6}$$

1.   We need to prove p(n) is true for n = 1

L.H.S. = $1^2$ = 1;

$$\text{R.H.S.} = \frac{1(2)(3)}{6} = 1$$

L.H.S. =  R.H.S.                    $\therefore$   p(n) is true for n = 1.

2.   We will assume p(n) is true for n = k

$$p(k): 1^2 + 2^2 + 3^2 + \ldots + k^2 = \frac{k(k+1)(2k+1)}{6} \tag{1}$$

3.   We need to prove p(n) is true for n = k + 1

$$p(k+1): 1^2 + 2^2 + 3^2 + \ldots + k^2 + (k+1)^2 = \frac{(k+1)(k+2)(2k+1)+1}{6}$$

$$\therefore 1^2 + 2^2 + 3^2 + \ldots + k^2 + (k+1)^2 = \frac{(k+1)(k+2)(2k+3)}{6}$$

$$\text{L.H.S.} = \underbrace{1^2 + 2^2 + 3^2 + \ldots + k^2} + (k+1)^2$$

$$\text{from (1)}$$

$$= \frac{k(k+1)(2k+1)}{6} + (k+1)^2$$

$$\text{L.H.S.} = (k+1)\left[\frac{k(2k+1)}{6} + (k+1)\right]$$

$$= \frac{(k+1)}{6}[2k^2 + k + 6k + 6]$$

$$= \frac{(k+1)[2k^2 + 7k + 6)}{6} \tag{2}$$

$$\text{R.H.S.} = \frac{(k+1)(k+2)(2k+3)}{6}$$

$$\frac{(k+1)[2k^2 + 3k + 4k + 6)}{6}$$

$$= \frac{(k+1)[2k^2 + 7k + 6]}{6} \tag{3}$$

$\therefore$ L.H.S.   =       R.H.S. from (2) and (3)

$\therefore$ p(n) is true for n = k + 1

$\therefore$ p(n) is true for all n ≥ 1

**Problem 2 :**

Show by mathematical induction $1^3 + 2^3 + ... + n^3 = \dfrac{n^3((n+1)^2}{6}$

$\Rightarrow$   Let p(n) :  $1^3 + 2^3 + ... + n^3 = \dfrac{n^3((n+1)^2}{4}$

(1)  We need to prove p(n) is true for n = 1

L.H.S.  = $1^3 = 1$ ;        R.H.S. $= \dfrac{1(1+1)^2}{4} = 1$

L.H.S.  =  R.H.S. $\therefore$  p(n) is true for n = 1.

(2)  We will assume p(k) is true for n = k

$p(k) : 1^3 + 2^3 + ... + k^3 = \dfrac{k^2(k+1)^2}{4}$

(3)  We need to prove p(n) is true for n = k + 1

$p(k+1) : 1^3 + 2^3 + ... + k^3 + (k+1)^3 = \dfrac{(k+1)^2(k+2)^2}{4}$

L.H.S. $= \underbrace{1^3 + 2^3 + ... + k^3} + (k+1)^3$

from (1)

$= \dfrac{k^2(k+1)^2}{4} + (k+1)^3$

L.H.S. $= (k+1)^2 \left[ \dfrac{k^2}{4} + (k+1) \right]$          $= \dfrac{(k+1)^2}{4}[k^2 + 4k + 4]$

$= \dfrac{(k+1)^2(k+2)^2}{4}$          $=$ R.H.S.

$\therefore$ L.H.S. $=$         R.H.S.

p(n) is true for n = k + 1

$\therefore$ p(n) is true for all n $\geq$ 1.

## 1.12. PRINCIPLE OF INCLUSION EXCLUSION

Consider that a task can be done in either m ways or n ways and let A represents a set with m elements and B represents a set with n elements. Also, both the sets have some elements in common. Sum rule will not work as we will end up adding common ways twice. This problem is called the double counting.

To avoid double counting, we need to add the count of all elements in set A and set B, and then subtract the count of common elements. This technique is called the principle of inclusion exclusion.

$$|A \cup B| = |A| + |B| - |A \cap B|$$

**Example 1 :**

How many integers from 1 to 20 are multiplies of 2 or 3 ?

**Solution :**

Let A represents a set containing multiples of 2 and B represents a set containing multiples of 3.

$$A = \{2, 4, 6, 8, 10, 12, 14, 16, 18, 20\}$$

$$B = \{3, 6, 9, 12, 15, 18\}$$

$$A \cap B = \{6, 12, 18\}$$

$$|A| = 10, |B| = 6 \text{ and } |A \cap B| = 3$$

$$|A \cup B| = |A| + |B| - |A \cap B|$$

$$|A \cup B| = 10 + 6 - 3$$

**Example 2 :**

How many bits strings of length eight either start with a 1 bit or end with two bits 0 0 ?

**Solution :**

Start with 1

$$\underline{1} \ \underbrace{\text{-------}}$$

$2^7 = 128$ ways

End with O O

$$\underbrace{\text{------}} \ \underline{O} \ \underline{O}$$

$2^6 = 64$ ways

Let A is a set containing all bit strings of length eight starting with 1 and B is a set containing all bit strings ending with O O.

$$|A| = 128 \ \text{and} \ |B| = 64$$

Strings in common

$$\underline{1} \ \underbrace{\text{--------}} \ \underline{O} \ \underline{O}$$

$2^5 = 32$ ways

$|A| = 128, |B| = 64 \ \text{and} \ |A \cap B| = 32$

According inclusion exclusion principle

$$|A \cup B| = |A| + |B| - |A \cap B|$$

$$|A \cup B| = 128 + 64 - 32 = 160$$

**Generally,**

Let A, B be any two finite sets.

Then $n (A \cup B) = n (A) + n (B) - n (A \cap B)$

Here "include" $n (A)$ and $n (B)$ and we "exclude" $n (A \cap B)$

Suppose A, B, C are finite sets.

Then $A \cup B \cup C$ is finite and

$$n(A \cup B \cup C) = n(A) + n(B) + n(C) - n(A \cap B) - n(A \cap C) - n(B \cap C) + n(A \cap B \cap C)$$

**Example:**

In a town of 100 families it was found that 40% of families buy newspaper A, 20% family buy newspaper B, 10% family buy newspaper C, 5% family buy newspaper A and B, 3% family buy newspaper B and C and 4% family buy newspaper A and C. If 2% family buy all the newspaper.

Find the number of families which buy

1.  Number of families which buy all three newspapers.
2.  Number of families which buy newspaper A only
3.  Number of families which buy newspaper B only
4.  Number of families which buy newspaper C only
5.  Number of families which buy None of A, B, C
6.  Number of families which buy exactly only one newspaper
7.  Number of families which buy newspaper A and B only
8.  Number of families which buy newspaper B and C only
9.  Number of families which buy newspaper C and A only
10. Number of families which buy at least two newspapers
11. Number of families which buy at most two newspapers
12. Number of families which buy exactly two newspapers

**Solution:**

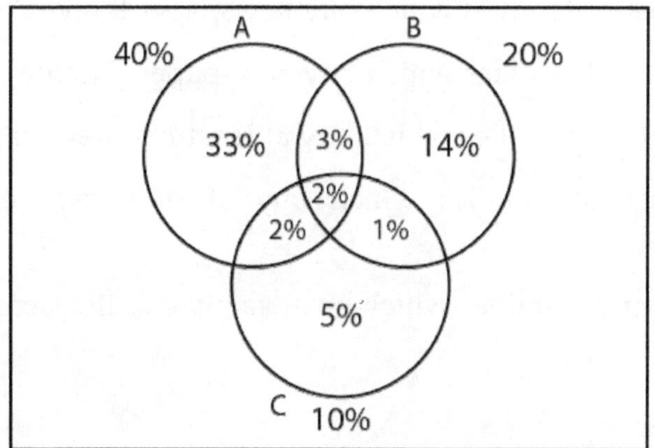



1.  Number of families which buy all three newspapers:

$$n(A \cup B \cup C) = n(A) + n(B) + n(C) - n(A \cap B) - n(A \cap C) - n(B \cap C) + n(A \cap B \cap C)$$

$$n(A \cup B \cup C) = 40 + 20 + 10 - 5 - 3 - 4 + 2 = 60\%$$

2.  Number of families which buy newspaper A only = 40 - 7 = 33%

3.  Number of families which buy newspaper B only = 20 - 6 = 14%

4.  Number of families which buy newspaper C only = 10 - 5 = 5%

5.  Number of families which buy None of A, B, and C

$$n(A \cup B \cup C)^c = 100 - n(A \cup B \cup C)$$

$$n(A \cup B \cup C)^c = 100 - [40 + 20 + 10 - 5 - 3 - 4 + 2]$$

$$n(A \cup B \cup C)^c = 100 - 60 = 40\%$$

6.  Number of families which buy exactly only one newspaper = 33 + 14 + 5 = 52%

7.  Number of families which buy newspaper A and B only = 3%

8.  Number of families which buy newspaper B and C only = 1%

9.  Number of families which buy newspaper C and A only = 2%

10. Number of families which buy at least two newspapers = 8%

11. Number of families which buy at most two newspapers = 98%

12. Number of families which buy exactly two newspapers = 6%

# Chapter 2

# GROWTH OF FUNCTIONS

## 2.1. ASYMPTOTIC NOTATIONS

Asymptotic Notations are used to represent the complexities of algorithms for asymptotic analysis. These notations are mathematical tools to represent the complexities. There are three notations that are commonly used. Asymptotic notations are based on the concept of functions. We need to be aware of the concept that which function is lesser or greater than which function before understanding the concept of Asymptotic notations.

The below is the precedence of the functions used normally in asymptotic notations to represent time complexities.

$$1 < \log n < n < n \log n < n^2 < n^3 < \ldots < 2^n < 3^n < \ldots < n^n$$

**1)  Big Oh Notation**

Big Oh (O) notation gives the upper bound for a function $f(n)$ within a constant factor.

The function $f(n) = O(g(n))$ if and only if there exists positive constants $c$ and $n_o$ such that $f(n) <= c * g(n)$ for all $n >= n_o$

This definition is represented by the below graph.

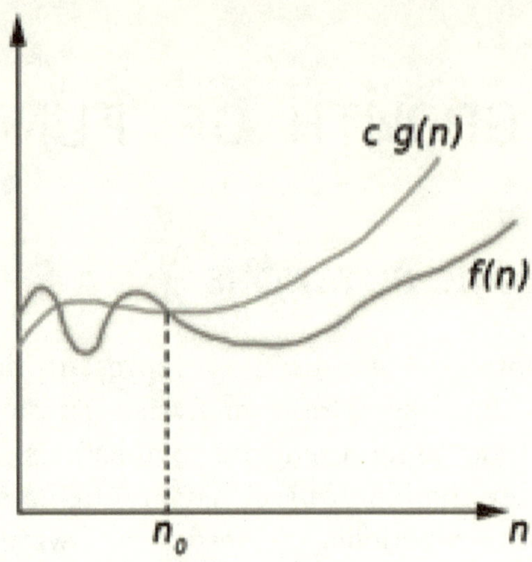


Example: f(n) = 2n + 3

We can say 2n + 3 <= 2n + 3n = 5n for all n >= 1

So, here f(n) = 2n + 3, c = 5 and g(n) = n as per the formula above.

Therefore f(n) = O(n) as g(n) = n

We can also say f(n) = $O(n^2)$ as Big Oh is the upper bound and we have seen that $n < n^2$

## 2)  Big Omega Notation

Big Omega () notation gives the lower bound for a function f(n) within a constant factor.

The function f(n) =  (g(n)) if and only if there exists positive constants c and $n_o$ such that f(n) >= c * g(n) for all n >= $n_o$

This definition is represented by the below graph.

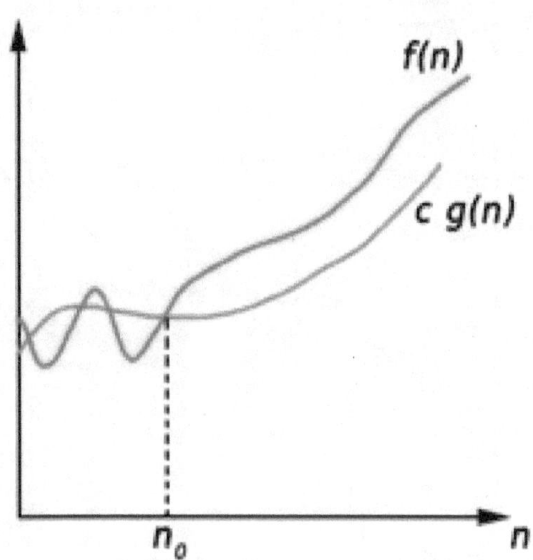


Example: $f(n) = 2n + 3$

We can say $2n + 3 >= 1n$ for all $n >= 1$

So, here $f(n) = 2n + 3$, $c = 1$ and $g(n) = n$ as per the formula above.

Therefore $f(n) = \Omega(n)$ as $g(n) = n$

We can also say $f(n) = \Omega(\log n)$ as Omega is the lower bound and we have seen that $n > \log n$

## 3)   Big Theta Notation

Big Theta ($\Theta$) notation gives the average bound for a function $f(n)$ within a constant factor.

The function $f(n) = \Theta(g(n))$ if and only if there exists positive constants $c_1$, $c_2$ and $n_o$ such that $c_1 * g(n) <= f(n) <= c_2 * g(n)$ for all $n >= n_o$

This definition is represented by the below graph.

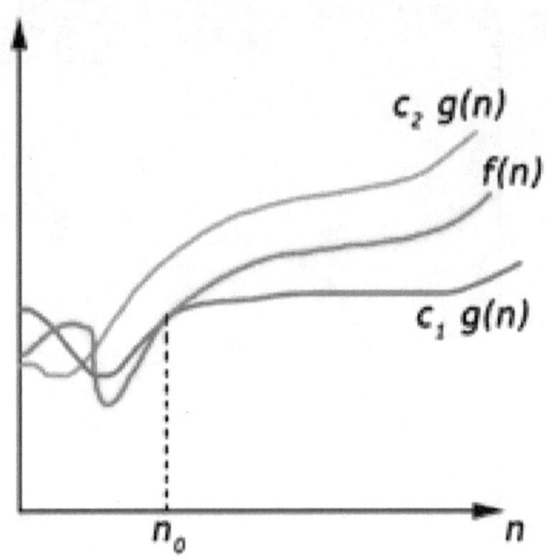


Example: f(n) = 2n + 3

We can say 1n <= 2n + 3 <= 5n for all n >= 1

So, here f(n) = 2n + 3, $c_1$ = 1, $c_2$ = 5 and g(n) = n as per the formula above.

Therefore f(n) =  (n) as g(n) = n

We can see more examples on this below.

## Example 1:

$$f(n) = 2n^2 + 3n + 4$$

We can say $2n^2 + 3n + 4 <= 2n^2 + 3n^2 + 4n^2 = 9n^2$ for all n >= 1

So, here $f(n) = 2n^2 + 3n + 4$, c = 9 and $g(n) = n^2$ as per the formula of Big Oh.

Therefore $f(n) = O(n^2)$ as $g(n) = n^2$

We can say $2n^2 + 3n + 4 >= 1n^2$ for all n >= 1

So, here $f(n) = 2n^2 + 3n + 4$, $c = 1$ and $g(n) = n^2$ as per the formula of Big Omega.

Therefore $f(n) = \ (n^2)$ as $g(n) = n^2$

We can say $1n^2 <= 2n^2 + 3n + 4 <= 9n^2$ for all $n >= 1$

So, here $f(n) = 2n^2 + 3n + 4$, $c_1 = 1$, $c_2 = 9$ and $g(n) = n^2$ as per the formula of Big Theta.

Therefore $f(n) = \ (n^2)$ as $g(n) = n^2$

**Example 2:**

$$f(n) = n^2 \log n + n$$

We can say $n^2 \log n + n <= 10\, n^2 \log n$ for all $n >= 1$

So, here $f(n) = n^2 \log n + n$, $c = 10$ and $g(n) = n^2 \log n$ as per the formula of Big Oh.

Therefore $f(n) = O(n^2 \log n)$ as $g(n) = n^2 \log n$

We can say $n^2 \log n + n >= 1n^2 \log n$ for all $n >= 1$

So, here $f(n) = n^2 \log n + n$, $c = 1$ and $g(n) = n^2 \log n$ as per the formula of Big Omega.

Therefore $f(n) = \ (n^2)$ as $g(n) = n^2 \log n$

We can say $1n^2 \log n <= n^2 \log n + n <= 10\, n^2 \log n$ for all $n >= 1$

So, here $f(n) = n^2 \log n + n$, $c_1=1$, $c_2=10$ and $g(n) = n^2 \log n$ as per the formula of Big Theta.

Therefore $f(n) = \ (n^2 \log n)$ as $g(n) = n^2 \log n$

## 2.2. SUMMATION FORMULAS AND PROPERTIES

### Summation Formulas

A summation formula, also known as a series formula, represents the sum of a sequence of numbers. It typically involves a concise notation to express the sum of terms in a sequence, often with a specific pattern or formulaic relationship between the terms.

Summation or sigma ($\Sigma$) notation is a method used to write out a long sum in a concise way. This notation can be attached to any formula or function. A summation is a short form of repetitive addition. We can also replace summation with a loop of addition.

For example, $\sum_{i=1}^{10}(i)$ is a sigma notation of the addition of finite sequence $1 + 2 + 3 + 4 \ldots \ldots + 10$ where the first element is 1 and the last element is 10.

Summation notation can be used in various fields of mathematics like Sequence in series,

Integration, Probability, Permutation and Combination, Statistics.

### Standard Summation Formulas

1.  Sum of First n Natural Numbers:

    $$(1+2+3+\ldots+n) = \sum_{i=1}^{n}(i) = [n \times (n+1)]/2$$

2.  Sum of Square of First n Natural Numbers:

    $$(1^2+2^2+3^2+\ldots+n^2) = \sum_{i=1}^{n}(i^2) = [n \times (n+1) \times (2n+1)]/6$$

3.  Sum of Cube of First n Natural Numbers:

    $$(1^3+2^3+3^3+\ldots+n^3) = \sum_{i=1}^{n}(i^3) = [n^2 \times (n+1)^2)]/4$$

4.  Sum of First n Even Natural Numbers:

$$(2+4+...+2n) = \sum_{i=1}^{n}(2i) = [n \times (n+1)]$$

5.  Sum of first n odd natural numbers:

$$(1+3+...+2n-1) = \sum_{i=1}^{n}(2i-1) = n^3$$

6.  Sum of Square of First n Even Natural Numbers:

$$(2^2+4^2+...+(2n)^2) = \sum_{i=1}^{n}(2i)^2 = [2n(n+1)(2n+1)]/3$$

7.  Sum of Square of First n Odd Natural Numbers:

$$(1^2+3^2+...+(2n-1)^2) = \sum_{i=1}^{n}(2i-1)^2 = [n(2n+1)(2n-1)]/3$$

8.  Sum of Cube of First n Even Natural Numbers:

$$(2^3+4^3+...+(2n)^3) = \sum_{i=1}^{n}(2i)^3 = 2[n(n+1)]^2$$

9.  Sum of Cube of First n Odd Natural Numbers:

$$(1^3+3^3+...+(2n-1)^3) = \sum_{i=1}^{n}(2i-1)^3 = n^2(2n^2-1)$$

10. For the Geometric Series a, ar, $ar^2$, $ar^3$, ... $ar^{n-1}$, the sum of first n terms is:

$$\sum_{i=1}^{n} ar^{i-1} = [a(1-r^n)]/(1-r)$$

$$\sum_{i=1}^{n} ar^{i-1} = a/(1-r), \text{when} |r|<1$$

## Summation Properties

Summation properties are rules and characteristics that govern the behavior of sums of sequences. These properties are essential for simplifying and manipulating summations in mathematics, particularly in areas like calculus, discrete mathematics, and algorithm analysis. Here are some important summation properties.

## Property 1

$$\sum_{i=1}^{n} c = c + c + c + \ldots + c(n) \text{ times} = nc$$

For example: Find the value of $\sum_{i=1}^{4} c$.

By using property 1 we can directly calculate the value of $\sum_{i=1}^{4} c$ as $4 \times c = 4c$.

## Property 2

$$\sum_{c=1}^{n} kc = (k \times 1) + (k \times 2) + (k \times 3) + \ldots + (k \times n)$$

$$\ldots (n) \text{ times} = k \times (1 + \ldots + n) = k \sum_{c=1}^{n} c$$

For example: Find the value of $\sum_{i=1}^{4} 5i$.

By using property 2 and 1 we can directly calculate value of $\sum_{i=1}^{4} 5i$ as

$$5 \times \sum_{i=1}^{4} i = 5 \times (1 + 2 + 3 + 4) = 50.$$

## Property 3

$$\sum_{c=1}^{n} (k \times c) = (k \times 1) + (k \times 2) + (k \times 3) + \ldots + (k \times n)$$

$$\ldots (n) \text{ times} = (n \times k) + (1 + \ldots + n) = nk + \sum_{c=1}^{n} c$$

For example: Find the value of $\sum_{i=1}^{4} (5 + i)$.

By using property 2 and 3 we can directly calculate value of $\sum_{i=1}^{4} (5 + i)$ as

$$5 \times 4 + \sum_{i=1}^{4} i = 20 + ((1 + 2 + 3 + 4) = 30.$$

## Property 4

$$\sum_{k=1}^{n} (f(k) + (g(k)) = \sum_{k=1}^{n} (f(k) + \sum_{k=1}^{n} g(k)$$

For example: Find value of $\sum_{i=1}^{4}(i+i^2)$.

By using property 4 we can directly calculate value of $\sum_{i=1}^{4}(i+i^2)$ as

$$\sum_{i=1}^{4} i + \sum_{i=1}^{4} i^2 = (1+2+3+4)+(1+4+9+16)=40.$$

## Example Problems using Summation Formulas and Properties

**Example 1:** Find the sum of first 10 natural numbers, using the summation formula.

**Solution:**

Using the summation formula for sum of n natural number

$$\sum_{i=1}^{n}(i) = [n \times (n+1)]/2$$

We have sum of first 10 natural numbers

$$\sum_{i=1}^{10}(i) = [10 \times (10+1)]/2 = 55$$

**Example 2:** Find the sum of 10 first natural numbers greater than 5, using the summation formula.

**Solution:**

Sum of 10 first natural numbers greater than

$$5 = \sum_{i=6}^{15}(i) = \sum_{i=1}^{15}(i) - \sum_{i=1}^{5}(i)$$

$$= [15 \times 16]/2 - [5 \times 6]/2 = 120 - 15 = 105$$

**Example 3:** Find the sum of given finite sequence

$$1^2 + 2^2 + 3^2 + \ldots 8^2.$$

**Solution:**

Given sequence is $1^2 + 2^2 + 3^2 + \ldots 8^2$, it can be written as the property/ formula of summation

$$\sum_{i=1}^{8} i^2 = [8 \times (8+1) \times (2 \times 8 + 1)] / 6 = [8 \times 9 \times 17] / 6 = 204$$

**Example 4:** Simplify $\sum_{c=1}^{n} kc$.

**Solution:**

Given summation formula

$$\sum_{c=1}^{n} kc = (k \times 1) + (k \times 2) + \ldots + (k \times n)(n \text{ terms})$$
$$= k\,(1 + 2 + 3 + \ldots + n) = \sum_{c=1}^{n} kc = k \sum_{c=1}^{n} c$$

**Example 5:** Simplify and evaluate $(4+x)$.

**Solution:**

Given summation is $\sum_{x=1}^{n}(4+x)$

As we know that $\sum_{c=1}^{n}(k \times c) = nk + \sum_{c=1}^{n} c$

Given summation can be simplified as, $4n + \sum_{x=1}^{n}(x)$

**Example 6:** Simplify $\sum_{x=1}^{n}(2x + x^2)$.

**Solution:**

Given summation is $\sum_{x=1}^{n}(2x + x^2)$.

as we know that $\sum_{k=1}^{n}(f(k) + (g(k)) = \sum_{k=1}^{n}(f(k) + \sum_{k=1}^{n} g(k)$

given summation can be simplified as $\sum_{x=1}^{n}(2x) + \sum_{x=1}^{n}(x^2)$.

## 2.3. BOUNDING SUMMATIONS

Bounding summations refer to the process of finding upper and lower bounds for the value of a summation or series. These bounds provide an approximation of the total value of the summation, often by bounding the terms within the series using simpler functions or inequalities.

Bounding summations can be useful in various mathematical and computational contexts, including algorithm analysis and estimating the performance of iterative processes. By bounding the sum of terms, one can gain insights into the behavior of the summation without needing to compute the exact value, which can be computationally expensive or infeasible for large series.

For example, in algorithm analysis, bounding the time complexity of an algorithm often involves bounding the number of operations performed in a loop or recursive function. This bounding helps in understanding how the algorithm scales with the size of the input, even if the exact number of operations cannot be determined easily.

In mathematical analysis, bounding summations can be used to estimate the value of infinite series or to prove convergence properties. By bounding the terms within the series, mathematicians can determine whether the series converges or diverges, and if it converges, they can estimate its value within a certain range.

## Techniques for Bounding Summations:

1) **Geometric Series Bounding:**

   Definition: Utilizes the formula for the sum of a geometric series to establish upper and lower bounds based on the initial term and common ratio.

2) **Telescoping Series:**

   Definition: Involves finding cancellations or simplifications within the series, resulting in a telescoping effect that aids in bounding.

**3) Integral Approximation:**

Definition: Replaces a discrete sum with a continuous integral, providing upper and lower bounds by comparing the series with a continuous approximation.

**4) Comparison Test:**

Definition: Compares the given series with a known series whose convergence or divergence behavior is well-understood to establish bounds.

**5) Inequalities:**

Definition: Utilizes mathematical inequalities to bound the terms of the series, providing constraints on their values.

**6) Cauchy Condensation Test:**

Definition: Specifically applicable to positive and decreasing series, involves comparing the original series with a condensed series to establish convergence or divergence.

**7) Abel's Summation Formula:**

Definition: Applies Abel's formula to relate the infinite series to the partial sums of the series, aiding in bounding.

**8) Majorization and Minorization:**

Definition: Finds sequences that majorize and minorize the series terms to establish convergence or divergence, providing upper and lower bounds.

**9) Limit Comparison Test:**

Definition: Compares the given series with another series, taking the ratio of their terms to determine convergence behavior and establish bounds.

**10) Ratio Test:**

Definition: Analyzes the convergence of a series by taking the ratio of consecutive terms, especially useful for series involving factorials or exponentials.

**11) Root Test:**

Definition: Examines the convergence of a series by taking the nth root of the absolute value of the terms, aiding in bounding.

**12) Bounding with Partial Sums:**

Definition: Uses partial sums of the series to provide upper and lower bounds for the entire series.

**13) Bounding with Integrals of Step Functions:**

Definition: Approximates the series by integrating step functions that closely resemble the series, aiding in bounding.

**14) Bounding with Mean Value Theorem:**

Definition: Utilizes the Mean Value Theorem for integrals to establish bounds for certain types of series.

**15) Bounding with Convexity:**

Definition: Leverages the convexity of certain functions to bound the terms of the series.

**16) Bounding with Convex Hulls:**

Definition: Uses the convex hull of a function to provide bounds for the series, especially useful in optimization problems.

**17) Bounding with Probabilistic Methods:**

Definition: Applies probabilistic techniques, such as Markov's inequality, to derive bounds for certain types of series.

**18) Bounding with Taylor Series:**

Definition: Uses Taylor series expansions to approximate the terms of the series and establish bounds.

**19) Bounding with Symmetry:**

Definition: Exploits symmetry properties of the series to simplify calculations and establish bounds.

**20) Bounding with Analytic Methods:**

Definition: Employs analytical methods, such as contour integration, to establish bounds for certain series in complex analysis.

Choosing the appropriate technique depends on the specific characteristics of the series being considered.

**Example 1:**

Bounding the Sum of a Finite Arithmetic Series:

Consider the arithmetic series:

$$\sum_{k=1}^{n} k = 1 + 2 + 3 + \ldots + n$$

We can bound this series by replacing each term with the largest and smallest possible values:

The largest term is n, occurring n times: $n + n + \ldots + n = n^2$

The smallest term is 1, occurring n times: $1 + 1 + \ldots + 1 = n$

Therefore, we have the bounds:

$$n \le \sum_{k=1}^{n} k \le n^2$$

These bounds give us an idea of the growth rate of the sum of the series.

**Example 2:**

Bounding a Geometric Series

Consider the geometric series $S=1+2+4+8+\ldots+2^n$.

The sum of this series is $S = 2^{n+1}-1$. Now, let's find an upper and lower bound without calculating the exact sum.

**Upper Bound:**

$S=2^{n+1}-1$ implies $S< 2^{n+1}$, so is upper-bounded by $2^{n+1}$.

**Lower Bound:**

$S=2^{n+1}-1$ implies $S>2^{n+1}-2^n$, so S is lower-bounded by $2^n$.

These bounds help in understanding that the sum S is between $2^n$ and $2^{n+1}$.

**Example 3:**

Bounding a Factorial Sum

Consider the series $S=1+1/1!+1/2!+1/3!+\ldots+1/n!$.

**Upper Bound:**

Observe that $S<1+1/1+1/2+1/2^2+\ldots=2$, so S is upper-bounded by 2.

**Lower Bound:**

Each term in the series is positive, so $S > 1$, and thus S is lower-bounded by 1.

This means $1< S< 2$, providing a range for the sum without calculating the exact value.

**Example 4:**

Comparison Test

Consider the series: $S = \sum_{n=1}^{\infty}(n^2+1)/(n^3+2)$

To determine whether this series converges or diverges, we can use the comparison test. Compare the given series to a simpler series whose behavior is known.

Let's consider the series $\sum_{n=1}^{\infty} 1/n$, which is a divergent harmonic series.

Since $(n^2+1)/(n^3+2)$ is always less than or equal to $1/n$ for all $n >= 1$, by the comparison test, if the harmonic series diverges, then the given series also diverges.

**Example 5:**

Bounding with Inequalities

Suppose you want to find a bound for the series: $S = \sum_{n=1}^{\infty}(n^2+1)/2^n$

To bound this series, observe that $(n^2+1)/2^n$ is less than or equal to $(n^2+n^2)/2^n = 2n^2/2^n$

Now we can use the ratio test on the series $\sum_{n=1}^{\infty} 2n^2/2^n$

The ratio test shows that this series converges since

$$\lim_{n \to \infty} a_{n+1}/a_n < 1.$$

Therefore, by the comparison test, the original series $\sum_{n=1}^{\infty}(n^2+1)/2^n$ also converges.

**Example 6:**

Cauchy Condensation Test

Consider the series: $S = \sum_{n=1}^{\infty} 1/n^2$

To determine convergence, we can use the Cauchy condensation test. The test states that if $a_n$ is a non-increasing sequence of positive terms, then $\sum_{n=1}^{\infty} a_n$ and $\sum_{n=0}^{\infty} 2^n a2^n$ either both converge or both diverge.

Applying the condensation test to the given series, we get $\sum_{n=0}^{\infty} 2^n (1/2^n)^2 = \sum_{n=0}^{\infty} 1/2^n$.

This is a convergent geometric series. Therefore, by the condensation test, the original series $\sum_{n=1}^{\infty} 1/n^2$ also converges.

## 2.4. APPROXIMATION BY INTEGRALS

Approximation by integrals, often referred to as numerical integration or quadrature, is a technique used to estimate the value of a definite integral when the integral cannot be evaluated analytically or when it's computationally impractical to do so.

The basic idea is to divide the interval of integration into smaller subintervals and approximate the function within each subinterval. There are several methods for numerical integration, including:

1) **Rectangular Rule:** This involves approximating the area under the curve using rectangles. The simplest version is the left rectangle rule, where the height of each rectangle is taken from the left endpoint of each subinterval.

2) **Trapezoidal Rule:** This method approximates the area under the curve using trapezoids. It's generally more accurate than the rectangular rule as it considers both endpoints of each subinterval.

3) **Simpson's Rule:** Simpson's rule approximates the curve using quadratic polynomials (parabolas) instead of straight lines. It tends to provide even more accurate results than the trapezoidal rule.

4) **Gaussian Quadrature:** This method uses specific weights and points within each subinterval to construct an approximation. It's often more accurate and efficient than the other methods but requires more computational effort.

5) **Composite Methods:** These methods involve breaking the interval of integration into smaller subintervals and applying one of the above techniques within each subinterval. The results from each subinterval are then combined to get an overall approximation.

Numerical integration methods are widely used in various fields such as physics, engineering, finance, and computer science where analytical solutions are not always feasible or practical to obtain. The choice of method depends on factors like the smoothness of the function, the required level of accuracy, and computational resources available.

**Definition:**

When a summation has the form $\sum_{k=m}^{n}(f(k)$, where f(k) is a monotonically increasing function, we can approximate it by integrals:

$$\int_{m-1}^{n} f(x)dx <= \sum_{k=m}^{n} f(k) <= \int_{m}^{n+1} f(x)dx$$

The below figures justifies this approximation. The summation is represented as the area of the rectangle in the figure, and the integral is shaded region under the curve.

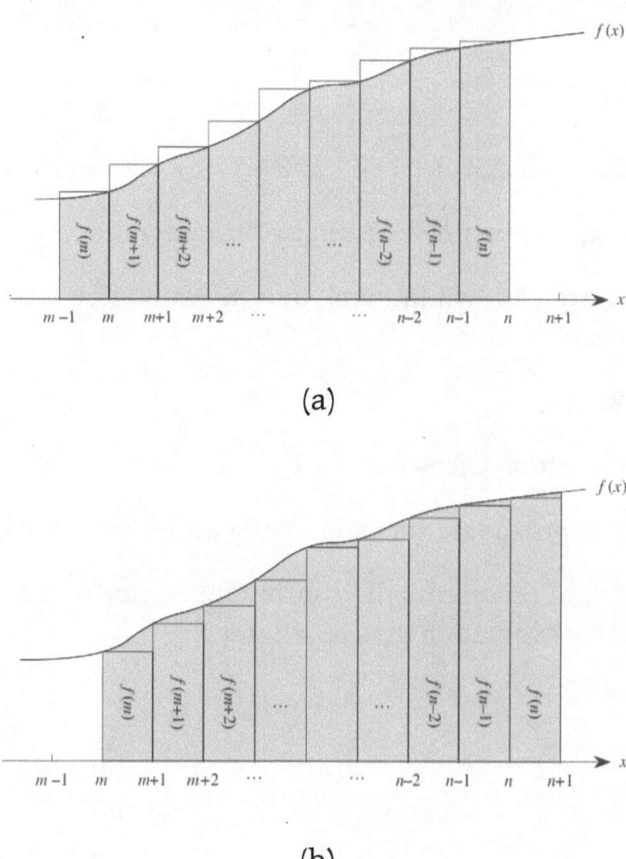

(a)

(b)

Approximation of $\sum_{k=m}^{n}(f(k)$ by integrals. The area of each rectangle is shown within the rectangle, and the total rectangle area represents the value of the summation. The integral is represented by the shaded area under the curve.

By comparing areas in (a), we get $\int_{m-1}^{n} f(x)dx <= \sum_{k=m}^{n} f(k)$, and then by shifting the rectangles one unit to the right, we get $\sum_{k=m}^{n} f(k) <= \int_{m}^{n+1} f(x)dx$.

When f(k) is monotonically decreasing function, we can use a similar method to provide the bounds.

$$\int_{m}^{n+1} f(x)dx <= \sum_{k=m}^{n} f(k) <= \int_{m-1}^{n} f(x)dx$$

**Example 1:**

Consider the series $S=1+1/2+1/3+1/4+$ ...

We can approximate this sum by an integral $S=\int_{1}^{\infty} \frac{1}{x} dx$

This is a well known integral, and its evaluation gives $S= \ln(x)$ evaluated from 1 to $\infty$.

**Example 2:**

Consider the geometric series $S=1+1/2+1/4+1/8+...$

We can approximate this sum using an integral $S=\int_{0}^{1} 2^{x}\, dx$

The integral represents the continuous analog of the geometric series, and its evaluation gives $S=2/\ln(2)$

# Chapter 3

# RECURRENCES

## 3.1. RECURRENCE RELATIONS

Recurrence relations, also known as recurrence equations, are mathematical equations that define sequences or functions recursively. Instead of directly specifying each term in the sequence or function, recurrence relations express each term in terms of previous terms in the sequence.

A recurrence relation typically consists of two parts:

**Base case:** This specifies the initial terms of the sequence or function, often for the first few terms, which serve as starting points for the recursion.

**Recurrence formula:** This defines how each subsequent term in the sequence or function depends on previous terms. It expresses the nth term in terms of one or more preceding terms.

Recurrence relations are commonly used in various fields of mathematics, computer science, physics, engineering, and more. They provide a powerful tool for describing processes that evolve over time or space, where the current state depends on previous states.

Examples of recurrence relations include:

**The Fibonacci sequence:**

$F(n) = F(n-1) + F(n-2)$, with base cases $F(0) = 0$ and $F(1) = 1$.

**The factorial function:** $n! = n \times (n-1)!$, with base case $0! = 1$.

**The Tower of Hanoi problem,** where the number of moves needed to solve the problem for n disks depends on the number of moves needed for $n-1$ disks.

Recurrence relations can often be solved analytically to find closed-form expressions for the sequence or function, but in many cases, they are solved numerically or through iterative methods. They are fundamental in algorithm design, dynamic programming, and the analysis of algorithms.

## Types of Recurrence Relations

Recurrence relations can be classified into several types based on their characteristics and properties. Some common types of recurrence relations include:

Linear recurrence relations: In linear recurrence relations, each term of the sequence is a linear combination of previous terms.

These relations are of the form: $a_n = c_1 \cdot a_{n-1} + c_2 \cdot a_{n-2} + \ldots + c_k \cdot a_{n-k} + f(n)$

where $c_1, c_2, \ldots, c_k$ are constants and $f(n)$ is a function of n.

**Homogeneous recurrence relations:** Homogeneous recurrence relations have no external function term ($f(n)=0$), meaning that the nth term of the sequence depends only on previous terms.

Example: Fibonacci sequence $F(n)=F(n-1)+F(n-2)$ is a homogeneous linear recurrence relation.

**Non-homogeneous recurrence relations:** Non-homogeneous recurrence relations have an external function term ($f(n) \neq 0$), meaning that the nth term of the sequence depends on both previous terms and an additional function of n.

Example: The recurrence relation $a_n = a_{n-1} + n$ is non-homogeneous because of the additional n term.

**First-order recurrence relations:** First-order recurrence relations involve only one previous term in defining the nth term of the sequence.

Example: The sequence defined by $a_n = a_{n-1} + 1$ is a first-order recurrence relation.

**Second-order recurrence relations:** Second-order recurrence relations involve two previous terms in defining the nth term of the sequence.

Example: The Fibonacci sequence $F(n) = F(n-1) + F(n-2)$ is a second-order recurrence relation.

**Constant-coefficient recurrence relations:** These are recurrence relations where the coefficients in the relation are constants.

Example: The recurrence relation $a_n = 2a_{n-1} - 3a_{n-2}$ is a constant-coefficient recurrence relation.

**Non-linear recurrence relations:** Non-linear recurrence relations involve terms that are not linearly dependent on previous terms.

Example: The recurrence relation $a_n = a_{n-1}^2$ is non-linear because the term $a_{n-1}^2$ is not linearly dependent on $a_{n-1}$.

These are some common types of recurrence relations, each with its own properties and methods for solving them. Depending on the specific recurrence relation, different techniques such as iteration, generating functions, characteristic equations, or matrix methods may be used to find solutions.

## 3.2. GENERATING FUNCTIONS

What is a numeric function ?

**Definition :**

A numeric function is a function defined from

$$W \rightarrow R \rightarrow \text{set of real numbers}$$

$$\downarrow \text{Set of whole numbers}$$

$$F : W \rightarrow R$$

**Examples :**

$$f(x) = x$$

$$f(x) = -x$$

$$f(x) = \sqrt{x}$$

$$f(x) = x^3$$

Let's take : $f(x) = x^3$

Domain $\{0, 1, 2, 3, 4, \ldots\}$

Codomain : R

$$f(0) = 0 \qquad f(1) = 1 \qquad f(2) = 8 \qquad f(3) = 27$$

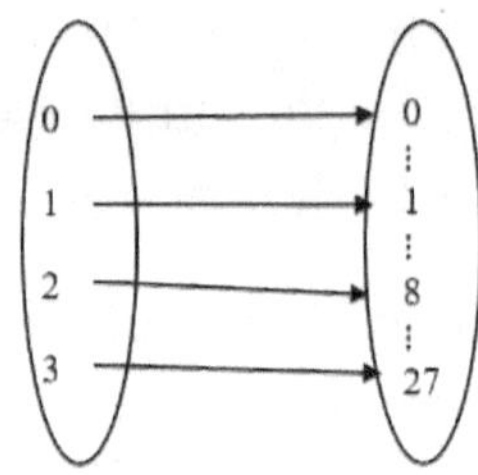

**Notation :**

Usually a numeric function is denoted by $a_r$

$$f(x) = x \equiv a_r = r$$

$$f(x) = -x \equiv a_r = -r$$

$$f(x) == a_r = \sqrt{r}$$

$$f(x) = x^3 \equiv a_r = r^3$$

**Determining sequence from a numeric function :**

What is a sequence ?

A sequence is a collection of objects in which repetitions are allowed and other does matter.

**Examples :**

$(0, 1, 2, 3, 4, 5, \ldots \}$ is a sequence

$(1, 1, 1, 1, 1, 1, \ldots \}$ is a sequence

**Consider the following numeric function :**

$$f(x) = x \quad [f : W \to R]$$

What is the sequence of the above function ?

**Solution :**

$$a_r = r$$

$$a_0 = 0$$

$$a_1 = 1$$

$$a_2 = 2$$

$$a_3 = 3$$

...

...

...

Range $(0, 1, 2, 3, 4, 5 \ldots \infty)$

Sequence : $(0, 1, 2, 3, 4, 5, \ldots \infty)$

$$f(x) = 1 \qquad [f : W \to R]$$

$$a_r = 1$$

$$a_0 = 1$$

$$a_1 = 1$$

$$a_2 = 1$$

$$a_3 = 1$$

...

...

...

Range $(1, 1, 1, 1, 1, \ldots \infty)$

Sequence : $(1, 1, 1, 1, 1, \ldots \infty)$

**Fact :**

For every numeric function, there is a generating function.

Let say $a_r$ is a numeric function.

If $a_r$ is a numeric function, then the generating function is represented by $G(x)$ where

$$G(x) = \sum_{r=0}^{\infty} a_r X^r$$

$$G(x) = a_0 X^0 + a_1 X^1 + a_2 X^2 + a_3 X^3 + \ldots a_r X^r + \ldots \infty$$

**Example 1 :**

Let say $a_r = (2, 6, 18, 54, 162, \ldots)$. Find the corresponding generating function.

**Solution :**

$$a_r = (2, 6, 18, 54, 162, \ldots).$$

$$G(x) = \sum_{r=0}^{\infty} a_r X^r$$

$$G(x) = a_0 X^0 + a_1 X^1 + a_2 X^2 + a_3 X^3 + \ldots + a_r X^r + \ldots \infty$$

$$G(x) = 2x^0 + 6x^1 + 18x^2 + 54x^3 + \ldots$$

$$= 2\,[1 + 3x + 9x^2 + 27x^3 + \ldots]$$

$$= 2\,[1 + (3x) + (3x)^2 + (3x)^3 + \ldots]$$

$$= 2(1-3x)^{-1} = \frac{2}{1-3x}$$

**Example 2 :**

What is the generating function for the sequence (1, 6, 36, 216, . . . ) ?

**Solution :**

$$G(x) = a^0 x^0 + a_1 x^1 + a_2 x^2 + a_3 x^3 + \ldots + a_r x^r + \ldots + \infty$$

$$G(x) = 1x^0 + 6x^1 + 36x^2 + 216x^3 + \ldots$$

$$= [1 + (6x) + (6x)^2 + (6x)^3 + \ldots]$$

$$= (1 - 6x)^{-1}$$

$$= \frac{1}{1-6x}$$

**Example 3 :**

What will be the sequence generated by the generating function $\dfrac{4x}{(1+x)^2}$ ?

**Solution :**

$$G(x) = \frac{4x}{(1+x)^2}$$

$$= 4x\,(1-x)^{-2}$$

$$= 4x[1 + 2x + 3x^2 + 4x^3 + \ldots]$$

$$= 4x + 8x^2 + 12x^3 + 16x^4 + \ldots$$

$$= 0x^0 + 4x^1 + 8x^2 + 12x^3 + 16x^4 + \ldots$$

$$\text{Sequence } (0, 4, 8, 12, 16, \ldots)$$

**What is standard formula ?**

Standard formula represents a standard generating function which can be used to derive other generating functions.

**Why standard formula ?**

Standard formula helps in finding the corresponding generating function quickly.

**Standard formula #1**

Let say the numeric function $a_r = 1$. Find the generating function.

We know that $G(x) = \sum_{r=0}^{\infty} a_r X^r$

As $a_r = 1$ $\qquad G(x) = \sum_{r=0}^{\infty} a_r X^r \quad 0 < x < 1$

$$\Rightarrow \quad x^0 + x^1 + x^2 + \ldots + x^\infty$$

$$\boxed{\begin{array}{l} \text{Let } x = 0.9 \\ (0.9)^0 \;=\; 1 \\ (0.9)^1 \;=\; 0.9 \\ (0.9)^2 \;=\; 0.81 \end{array}}$$

$$\text{This is GP} = \left(\frac{1-x^\infty}{1-x}\right) = \frac{1}{1-x}$$

## Geometric Progression – Review

General form of GP : $ar^0, ar^1, ar^2, ar^3 \ldots$

$$S_\infty = a + ar + ar^2 + ar^3 + \ldots ar^\infty$$

$$= a\left(\frac{1-r^\infty}{1-r}\right)$$

$$\text{If } a_r = 1, \text{then GF} = \frac{1}{1-x}$$

## Standard Formula #1

If $a_r = 1$, then $GF = \dfrac{1}{1-x}$

## Variation (Example 1)

What will be the generating function when $a_r = 5$.

We know that, $G(x) = \sum_{r=0}^{\infty} a_r X^r$

$$= \sum_{r=0}^{\infty} 5x^r \;=\; 5 \sum_{r=0}^{\infty} x^r$$

$$= 5\frac{1}{1-x} = \frac{5}{1-x}$$

So generating function for $(5, 5, 5, 5, \ldots, \infty)$ is $\dfrac{5}{1-x}$

## Variation (Example 2)

What will be the generating function when $a_r = 10$.

We know that, $G(x) = \sum_{r=0}^{\infty} a_r X^r$

$$= \sum_{r=0}^{\infty} 10x^r \;=\; 10 \sum_{r=0}^{\infty} x^r$$

$$= 10\frac{1}{1-x} = \frac{10}{1-x}$$

So generating function for $(10, 10, 10, 10, \ldots, \infty)$ is $\dfrac{10}{1-x}$

## Finding Coefficients (Example 1)

If $G(x) = \dfrac{1}{1-x}$ then what is $a_{15}$ ?

If $a_r = 1$, then GF $= \dfrac{1}{1-x}$

As $a_r = 1$, sequence must be $(1, \quad 1, \quad 1, \quad 1, \ldots)$

$$\uparrow \quad \uparrow \quad \uparrow \quad \uparrow$$
$$a_0 \quad a_1 \quad a_2 \quad a_3$$

Therefore, $a_{15} = 1$

## Finding Coefficients (Example 2)

If $G(x) = \dfrac{1}{1-x}$ then what is the coefficient of $x^{30}$?

$$\sum_{r=0}^{\infty} X^r = \frac{1}{1-x}$$

$$1 + x + x^2 + x^3 + x^4 + \ldots$$

all coefficients are 1.

Therefore coefficient of $x^{30} = 1$

## Standard formula # 2

$$\text{If } G(x) = \frac{1}{1-x} \text{ then } a_r = (-1)^r$$

## Proof :

We know that $\sum_{r=0}^{\infty} X^r = \dfrac{1}{1-x}$

Replace x by $-x$

$$\sum_{r=0}^{\infty} (-x)^r = \frac{1}{1+x} = \sum_{r=0}^{\infty} (-1)^r x^r$$

Therefore, $a_r = (-1)^r$

**Sequence :**

$$(1, -1, 1, -1, 1, \dots)$$

**Closed Form :**

$$G(x) = \frac{1}{1-x}$$

**Open Form :**

$$G(x) = 1 - x + x^2 - x^3 + x^4 + \dots$$

**Standard Formula #3**

$$\text{If } G(x) = \frac{1}{1-kx} \text{ then } a_r = k^r$$

**Proof :**

We know that $\sum_{r=0}^{\infty} x^r = \frac{1}{1-x}$

Replace x by kx

$$\sum_{r=0}^{\infty} (kx)^r = \frac{1}{1-kx} = \sum_{r=0}^{\infty} k^r x^r$$

Therefore, $a_r = k^r$

**Sequence :**

$$(1, k, k^2, k^3, k^4, \dots)$$

**Closed Form :**

$$G(x) = \frac{1}{1-kx}$$

**Open Form :**

$$G(x) = 1 + kx + k^2 x^2 + \ldots$$

**Standard Formula #4**

If $G(x) = \dfrac{1}{1 - kx}$ then $a_r = (-k)^r$

**Proof :**

We know that $\sum_{r=0}^{\infty} x^r = \dfrac{1}{1-x}$

Replace x by $-kr$

$$\sum_{r=0}^{\infty} (-kx)^r = \frac{1}{1+kx} = \sum_{r=0}^{\infty} (-k)^r x^r$$

Therefore, $a_r = (-k)^r$

**Sequence :**

$$(1, -k, k^2, -k^3, \ldots)$$

**Closed Form :**

$$G(x) = \frac{1}{1 + kx}$$

**Open Form :**

$$G(x) = 1 - kx + k^2 x^2 - \ldots$$

**Problem 1 :**

What is the corresponding numeric function for the generating function G(x) given below : $G(x) = \dfrac{1}{1 + 2x}$

**Solution :**

We know that if $G(x) = \dfrac{1}{1 - kx}$ then $a_r = k^r$.

Here k = 2

Therefore, $a_r = 2^r$

Sequence : (1, 2, 4, 8, 16, 32, . . . )

## Problem 2 :

What is the corresponding numeric function for the generating function G(x) given below : $G(x) = \dfrac{1}{1+5x}$

## Solution :

We know that if $G(x) = \dfrac{1}{1+kx}$ then $a_r = (-k)^r$

Here, k = 5

Therefore, $a_r = (-5)^r$

Sequence : (1, -5, 25, -125, . . . )

Coefficient of $x^4$ is $a_4 = 625$.

## Standard Formula #5

If $G(x) = \dfrac{k}{1-x}$ then $a_r = k$

## Proof :

We know that $\sum_{r=0}^{\infty} x^r = \dfrac{1}{1-x}$      for $a_r = 1$

Multiply G(x) by k

$k\sum_{r=0}^{\infty} x^r = \dfrac{k}{1-x} = \sum_{r=0}^{\infty} kx^r$

Therefore, $a_r = k$

**Sequence :**

$$(k, k, k, k, k, \ldots)$$

**Closed Form :**

$$G(x) = \frac{k}{1-x}$$

**Open Form :**

$$G(x) = k + kx + kx^2 + \ldots$$

**Remember that**

If $G(x)$ is the generating function for the numeric function $a_r$

$$a_r \rightarrow G(x)$$

$$ka_r \rightarrow k\,G(x)$$

**Standard formula # 6 :**

$$\text{If } G(x) = \frac{k}{1-x} \text{ then } a_r = (-1)^r.k$$

**Proof :**

We know that $\sum_{r=0}^{\infty}(-1)^r x^r = \dfrac{1}{1+x}$

$$a_r = (-1)^r$$

Multiply $G(x)$ by $k$,

$$k\sum_{r=0}^{\infty}(-1)^r x^r = \frac{k}{1-kx} = \sum_{r=0}^{\infty} k(-1)^r x^r$$

Therefore, $a_r = k\,(-1)^r$

**Sequence :**

$$(k, -k, k, -k, k, \ldots)$$

**Closed Form :**

$$G(x) = \frac{k}{1+x}$$

**Open Form :**

$$G(x) = k - kx + kx^2 - \ldots$$

In general,

$$\text{If } G(x) = \frac{k}{1-zx} \text{ then } a_r = k.(-Z)^r$$

**Proof :**

We know that $\sum_{r=0}^{\infty}(-1)^r x^r = \dfrac{1}{1+x}$ $a_r = (-1)^r$

Multiply $G(x)$ by $k$ and replace by $Zx$

$$k\sum_{r=0}^{\infty}(-1)^r(Zx)^r = \frac{1}{1+Zx} = \sum_{r=0}^{\infty} k(-Z)^r x^r$$

Therefore, $a_r = k \cdot (-Z)^r$

**Important Results**

$$\text{If } G(x) = \frac{k}{1-zx} \text{ then } a_r = k.(-Z)^r$$

$$\text{If } G(x) = \frac{k}{1-zx}, \text{ then } a_r = k.Z^r$$

**Problem 1 :**

What is the corresponding numeric function for the generating function $G(x)$ given below.

$$G(x) = \frac{5}{1-2x}$$

**Solution :**

We know that if $G(x) = \dfrac{k}{1+zx}$ then $a_r = k.Z^r$

Here, k = 5, z = 2

Therefore, $a_r$ = 5 . $2^r$

Sequence : (5, 10, 20, 40, 80, . . . )

**Problem 2 :**

What is the corresponding numeric function for the generating function G(x) given below.

$$G(x) = \dfrac{10}{1+7x}$$

Also determine the sequence

## Solution :

We know that if $G(x) = \dfrac{k}{1+zx}$ then $a_r = k.(-Z)^r$

Here, k = 10, Z = 7

Therefore, $a_r$ = 10 . $(-7)^r$

Sequence : (10, – 70, 490, . . . )

**Standard Formula # 7 :**

$$\text{If } G(x) = \dfrac{1}{a-x} \text{ then } a_r = \left(\dfrac{1}{a}\right)^{r+1}$$

**Proof :**

We know that $\sum_{r=0}^{\infty} x^r = \dfrac{1}{1-x} \qquad a_r = 1$

Multiplying the above generating function by $\frac{1}{a}$ and divide x by a.

$$\frac{\frac{1}{a}}{1-\frac{x}{a}} = \sum_{r=0}^{\infty} \frac{1}{a}\left(\frac{x}{a}\right)^r = \sum_{r=0}^{\infty} \left(\frac{1}{a}\right)^{r+1} x^r$$

**Therefore,** $a_r = \left(\frac{1}{a}\right)^{r+1}$

**Standard Formula # 8 :**

If $G(x) = \dfrac{1}{a+x}$ then $a_r = \dfrac{1}{a}\left(\dfrac{-1}{a}\right)^r$

**Proof :**

We know that $\sum_{r=0}^{\infty} x^r = \dfrac{1}{1-x}$ $\qquad a_r = 1$

Multiplying the above generating function by $\frac{1}{a}$ and divide x by $-a$.

$$\frac{\frac{1}{a}}{1-\frac{x}{a}} = \sum_{r=0}^{\infty} \frac{1}{a}\left(\frac{-x}{a}\right)^r = \sum_{r=0}^{\infty} \frac{1}{a}\left(\frac{-1}{a}\right)^r x^r$$

Therefore, $a_r = \dfrac{1}{a}\left(\dfrac{-1}{a}\right)^r$

**Problem 1 :**

What is the corresponding numeric function for the generating function G(x) given below.

$$G(x) = \frac{10}{10+2x}$$

**Solution :**

Multiply numerator and denominator by $\dfrac{1}{10}$

$$\dfrac{\dfrac{10}{10}}{\dfrac{10}{10}-\dfrac{2x}{10}}=\dfrac{1}{1-\dfrac{x}{5}} \qquad a_r=\left(\dfrac{1}{5}\right)^r$$

Coefficient of $x^{50}$.

**Solution :**

$$a_r=\left(\dfrac{1}{5}\right)^r$$

$$a_{50}=\left(\dfrac{1}{5}\right)^{50}$$

**Problem 2 :**

What is the corresponding numeric function for the generating function G(x) given below :

$$G(x)=\dfrac{6}{7+5x}$$

**Solution :**

Multiply numerator and denominator by $\dfrac{1}{7}$

$$\dfrac{\dfrac{6}{7}}{\dfrac{7}{7}+\dfrac{5x}{7}}=\dfrac{\dfrac{6}{7}}{1+\dfrac{5x}{7}} \qquad a_r=\dfrac{6}{7}\left(\dfrac{-5}{7}\right)^r$$

## Important properties

$$G(x) \rightarrow a_r$$
$$k.G(x) \rightarrow k.a_r$$

$$G(x) \rightarrow a_r$$
$$G(kx) \rightarrow k^r.a_r$$

## Standard Formula # 9 :

$$\text{If } G(x) = \frac{1}{1-x^2} \text{ then } a_r = \{^{0\,|\,r\,=\,2i\ i\geq0}_{0\,|\,r\,\neq\,2i}$$

## Proof :

We know that $\sum_{r=0}^{\infty} x^r = \dfrac{1}{1-x}$     $a_r = 1$

Replace x by $x^2$

$$\sum_{r=0}^{\infty} x^{2r} = \frac{1}{1-x^2}$$

## Open Form :

$$1 + x^2 + x^4 + x^6 + \ldots$$

## Sequence :

$$(1, 0, 1, 0, 1, 0, 1, 0, \ldots)$$

$$a_r = \begin{cases} 1\,|\,r = 2i\ \ i \geq 0 \\ 0\,|\,r \neq 2i \end{cases}$$

## Standard Formula # 10 :

$$\text{If } G(x) = \frac{1}{1-x^3} \text{ then } a_r = \{^{0\,|\,r\,=\,3i\ i\geq0}_{0\,|\,r\,\neq\,3i}$$

**Proof :**

We know that $\sum_{r=0}^{\infty} x^r = \dfrac{1}{1-x}$ $\qquad a_r = 1$

Replace x by $x^3$

$$\sum_{r=0}^{\infty} x^{3r} = \dfrac{1}{1-x^3}$$

**Open Form :**

$$1 + x^3 + x^6 + x^9 + \ldots$$

**Sequence :**

$$(1, 0, 0, 1, 0, 0, 1, 0, 0, 1, \ldots)$$

$$a_r = \begin{cases} 1 \mid r = 3i \;\; i \geq 0 \\ 0 \mid r \neq 3i \end{cases}$$

In general,

If $G(x) = \dfrac{1}{1-x^k}$ then $a_r = \begin{cases} 1 \mid r = ki \;\; i \geq 0 \\ 0 \mid r \neq ki \end{cases}$

**Proof :**

We know that $\sum_{r=0}^{\infty} x^r = \dfrac{1}{1-x}$ $\qquad a_r = 1$

Replaced x by $x^k$

$$\sum_{r=0}^{\infty} x^{kr} = \dfrac{1}{1-x^k}$$

**Open Form :**

$$1 + x^k + x^{2k} + x^{3k} + \ldots$$

**Sequence :**

$$(1, 0, 0, \ldots, 1, 0, 0, \ldots, 1, 0, 0, \ldots)$$

k – 1 zeros

$$a_r = \begin{cases} 1 \mid r = ki & i \geq 0 \\ 10 \mid r \neq ki \end{cases}$$

## Problem 1 :

What is the corresponding numeric function for the generating function $G(x)$ given below :

$$G(x) = \frac{9}{1-x^4}$$

## Solution :

Consider that if $G(x) = \dfrac{1}{1-x^4}$

**Sequence :**

$(1, 0, 0, 0, 1, 0, 0, 0, \ldots)$

$$a_r = \begin{cases} 1 \mid r = 4i & i \geq 0 \\ 0 \mid r \neq 4i \end{cases}$$

## Open Form :

$9 + 9x^4 + 9x^8 + \ldots$

## Sequence :

$(9, 0, 0, 0, 9, 0, 0, 0, \ldots)$

If $G(x) = \dfrac{9}{1-x^4}$

$$a_r = \begin{cases} a \mid r = 4i & i \geq 0 \\ 0 \mid r \neq 4i \end{cases}$$

## Problem 2 :

What is the corresponding numeric function for the generating function G(x) given below :

$$G(x) = \frac{7}{1-(3x)^4}$$

### Solution :

Consider that if $G(x) = \dfrac{1}{1-x^4}$

### Sequence :

$$(1, 0, 0, 0, 1, 0, 0, 0, \ldots)$$

$$a_r = \begin{cases} 1 & | \ r = 4i \quad i \geq 0 \\ 0 & | \ r \neq 4i \end{cases}$$

### Open Form :

$$7 + 7x^4 + 7x^8 + \ldots$$

### Sequence :

$$(7, 0, 0, 0, 7, 0, 0, 0, \ldots)$$

$$\text{If } G(x) = \frac{7}{1-x^4}$$

$$a_r = \begin{cases} 7 & | \ r = 4i \quad i \geq 0 \\ 0 & | \ r \neq 4i \end{cases}$$

$$\text{If } G(x) = \frac{7}{1-(3x)^4}$$

$$a_r = \begin{cases} 7 \times 3^r & | \ r = 4i \quad i \geq 0 \\ 0 & \quad \ \ | \ r \neq 4i \end{cases}$$

## Important properties

$$G(x) \to a_r$$
$$k.G(x) \to k.a_r$$

$$G(x) \to a_r$$
$$G(kx) \to k^r.a_r$$

## In general :

$$\text{If } G(x) = \frac{k}{1-(zx)^m} \text{ then}$$

$$a_r = \begin{cases} k \times Z^r & | \, r = mi \quad i \geq 0 \\ 0 & | \, r \neq mi \end{cases}$$

## Hence, generating function is represented by

$$G(x) = a_0 + a_1x + a_2x^2 + a_3x^3 + \ldots.$$

## The below formulas will be useful in solving generating functions:

$$(1 + x)^{-1} = 1 - x + x^2 - x^3 + \ldots$$
$$(1 - x)^{-1} = 1 + x + x^2 + x^3 + \ldots$$
$$(1 + x)^{-2} = 1 - 2x + 3x^2 - 4x^3 + \ldots$$
$$(1 - x)^{-2} = 1 + 2x + 3x^2 + 4x^3 + \ldots$$
$$(x^5 - 1) / (x - 1) = x^4 + x^3 + x^2 + x + 1$$
$$(x^5 + 1) / (x + 1) = x^4 - x^3 + x^2 - x + 1$$

## Find generating functions for the following:

1) Sequence: 1, 1, 1, ...

   Generating Function is $G(x) = 1 + x + x^2 + \ldots$

   That is $G(x) = (1 - x)^{-1}$

2) Sequence: 1, 1, 1, 1, 1, 1, 0

Generating Function is $G(x) = 1 + x + x^2 + x^3 + x^4 + x^5$

That is $G(x) = (x^6 - 1) / (x - 1)$

3) Sequence: 1, 3, 3, 1, 0, 0

Generating Function is $G(x) = 1 + 3x + 3x^2 + x^3$

That is $G(x) = (1 + x)^3$

4) Sequence: 1, a, $a^2$, $a^3$, ...

Generating Function is $G(x) = 1 + ax + a^2x^2 + a^3x^3 + ...$

That is $G(x) = (1 - ax)^{-1}$

5) Sequence: 1, -3, 9, -27, ...

Generating Function is $G(x) = 1 - 3x + 9x^2 - 27x^3 + ...$

That is $G(x) = (1 + 3x)^{-1}$

6) Sequence: 1, 2, 3, 4, ...

Generating Function is $G(x) = 1 + 2x + 3x^2 + 4x^3 + ...$

That is $G(x) = (1 - x)^{-2} = 1 / (1 - x)^2$

7) Sequence: 0, 1, 2, 3, ...

Generating Function is $G(x) = x + 2x^2 + 3x^3 + ...$

That is $G(x) = x (1 + 2x + 3x^2 + ...) = x (1 - x)^{-2} = x / (1 - x)^2$

8) Sequence: 0, 0, 0, 1, 2, 3, ...

Generating Function is $G(x) = x^3 + 2x^4 + 3x^5 + ...$

That is $G(x) = x^3 (1 + 2x + 3x^2 + ...) = x^3 (1 - x)^{-2} = x^3/(1 - x)^2$

9) Sequence: 0, 0, 0, 1, 1, 1, ...

Generating Function is $G(x) = x^3 + x^4 + x^5 + ...$

That is $G(x) = x^3 (1 + x + x^2 + ...) = x^3 (1 - x)^{-1} = x^3 / (1 - x)$

10)  Sequence: 1, -2, 3, -4, 5,  ...

Generating Function is $G(x) = 1 - 2x + 3x^2 - 4x^3 + ...$

That is $G(x) = (1 + x)^{-2}$

11)  Sequence: 1, 1, 0, 1, 1, 1, ...

Generating Function is $G(x) = 1 + x + x^3 + x^4 + x^5 + ...$

That is $G(x) = (1 - x)^{-1} - x^2$

12)  Sequence: 1, 0, -1, 0, 1, 0, -1, ...

Generating Function is $G(x) = 1 - x^2 + x^4 - x^6 + ...$

That is $G(x) = (1 + x^2)^{-1}$

13)  Sequence: 2, -2, 2, -2, 2, -2, ...

Generating Function is $G(x) = 2 - 2x + 2x^2 - 2x^3 + 2x^4 - 2x^5 + ...$

That is $G(x) = 2(1 - x + x^2 - x^3 + x^4 - x^5 + ....) = 2(1 + x)^{-1} = 2/(1 + x)$

## 3.3.  LINEAR RECURRENCE RELATIONS WITH CONSTANT COEFFICIENTS

Linear Recurrence Relations with Constant Coefficients are of the form:

$$a_n = c_1 {}^* a_{n-1} + c_2 {}^* a_{n-2} + ... + c_k {}^* a_{n-k} + f(n)$$

where $c_1, c_2, ..., c_k$ are constants and $f(n)$ is a function of n.

- The solution for this type of recurrence relation when the function $f(n)$ is given is:

$$a_n = a_n(h) + a_n(p)$$

- The solution for this type of recurrence relation when the function $f(n)$ is not given is:

$$a_n = a_n(h)$$

**To find $a_n(h)$:**

**Two roots:**

**1.  Different roots**

$$a_n(h) = u(S_1)^n + v(S_2)^n$$

When there are two roots and they are different roots, $a_n(h) = u(S_1)^n + v(S_2)^n$, where $S_1$ and $S_2$ are the two roots and u and v are constants.

**2.  Same roots**

$$a_n(h) = u(S)^n + v(S)^n$$

When there are two roots and they are same, $a_n(h) = u(S)^n + v(S)^n$, where S is the root and u and v are constants.

**Three roots:**

**1.  Different roots**

$$a_n(h) = u(S_1)^n + v(S_2)^n + w(S_3)^n$$

When there are three roots and they are different roots, $a_n(h) = u(S_1)^n + v(S_2)^n + w(S_3)^n$, where $S_1$, $S_2$ and $S_3$ are the three roots and u, v and w are constants.

**2.  Same roots**

$$a_n(h) = u(S)^n + vn(S)^n + wn^2(S)^n$$

When there are three roots and they are same, $a_n(h) = u(S)^n + vn(S)^n + wn^2(S)^n$, where S is the root and u, v and w are constants.

**To find $a_n(p)$:**

If $f(n) = r^n$ then $a_n(p) = Ar^n$

If $f(n) = c$, then $a_n(p) = A$

**Example 1:**

Solve the following recurrence relation $a_n = 4a_{n-1} + 5a_{n-2}$, where $a_1 = 2$, $a_2 = 6$

**Solution:**

Recall, Linear Recurrence Relations with Constant Coefficients are of the form:

$$a_n = c_1 {}^* a_{n-1} + c_2 {}^* a_{n-2} + \ldots + c_k {}^* a_{n-k} + f(n)$$

where $c_1, c_2, \ldots, c_k$ are constants and $f(n)$ is a function of n.

So, in the given example, $f(n)$ is not present and hence $a_n = a_n(h)$

In the given problem, we have

$$a_n = 4a_{n-1} + 5a_{n-2}$$

This can be written as

$$a_n - 4a_{n-1} - 5a_{n-2} = 0$$

The same can be written in quadratic form of equation as

$$x^2 - 4x^1 - 5x^0 = 0$$

$$x^2 - 4x - 5 = 0$$

The roots for the quadratic equation are $x = 5$, and $x = -1$

So, we have two different roots and hence the

$$a_n(h) = u(S_1)^n + v(S_2)^n$$

As we have in this case $a_n = a_n(h)$,

$$a_n = u(S_1)^n + v(S_2)^n$$

Here, $S_1 = 5$ and $S_2 = -1$

$$a_n = u(5)^n + v(-1)^n \qquad (1)$$

As, we have $a_1 = 2$ and $a_2 = 6$ in the given problem,

put $n = 1$ in (1)

$$a_1 = u(5)^1 + v(-1)^1 \quad ==> \quad a_1 = 5u - v$$

$$2 = 5u - v \qquad (2)$$

put $n = 2$ in (1)

$$a_2 = u(5)^2 + v(-1)^2 \quad ==> \quad a_2 = 25u + v$$

$$6 = 25u + v \qquad (3)$$

Solving the equations (2) and (3) simultaneously, we get

$$u = 4/15 \text{ and } v = -2/3$$

Substituting, these values of u and v in (1), we get

$$a_n = 4/15\ (5)^n - 2/3\ (-1)^n$$

Hence, this is the solution for the given recurrence relation.

**Example 2:**

Solve the following recurrence relation $a_n = 4a_{n-1} + 5a_{n-2} + 7^n$

**Solution:**

Recall, Linear Recurrence Relations with Constant Coefficients are of the form:

$$a_n = c_1{}^*a_{n-1} + c_2{}^*a_{n-2} + \ldots + c_k{}^*a_{n-k} + f(n)$$

where $c_1, c_2, \ldots, c_k$ are constants and $f(n)$ is a function of n.

So, in the given example, $f(n)$ is present and it is $7^n$ and hence $a_n = a_n(h) + a_n(p)$

In the given problem, we have

$$a_n = 4a_{n-1} + 5a_{n-2} + 7^n$$

This can be written as

$$a_n - 4a_{n-1} - 5a_{n-2} = 7^n \qquad (1)$$

The same can be written in quadratic form of equation as

$$x^2 - 4x^1 - 5x^0 = 0$$

$$x^2 - 4x - 5 = 0$$

The roots for the quadratic equation are $x = 5$, and $x = -1$

So, we have two different roots and hence the

$$a_n(h) = u(S_1)^n + v(S_2)^n$$

Here, $S_1 = 5$ and $S_2 = -1$

$$a_n(h) = u(5)^n + v(-1)^n \qquad (2)$$

Recall, to find $a_n(p)$, If $f(n) = r^n$ then $a_n(p) = Ar^n$

So, we have

$$a_n(p) = A7^n \qquad (3)$$

Put (3), $a_n = A7^n$ in (1), we get

$$A7^n - 4*A7^{n-1} - 5*A7^{n-2} = 7^n$$

$$A7^n - 4A*7^n/7 - 5A * 7^n/7^2 = 7^n$$

Canceling $7^n$ throughout the equation, we have,

$$A - 4/7A - 5/49A = 1$$

$$A[1 - 4/7 - 5/49] = 1 \quad ==> \quad A[16/49] = 1$$
$$A = 49/16 \tag{4}$$

Put (4) in (3), we get

$$a_n(p) = 49/16 * 7^n \tag{5}$$

From (2) and (5), we have

$$a_n = u(5)^n + v(-1)^n + 49/16 * 7^n$$

Hence, this is the solution for the given recurrence relation.

### Example 3:

Solve the following recurrence relation $a_n = 4a_{n-1} + 5a_{n-2} + 7$, where $a_1 = 2$ and $a_2 = 6$

### Solution:

Recall, Linear Recurrence Relations with Constant Coefficients are of the form:

$$a_n = c_1 * a_{n-1} + c_2 * a_{n-2} + \ldots + c_k * a_{n-k} + f(n)$$

where $c_1, c_2, \ldots, c_k$ are constants and $f(n)$ is a function of n.

So, in the given example, $f(n)$ is present and it is 7 and hence $a_n = a_n(h) + a_n(p)$

In the given problem, we have $a_n = 4a_{n-1} + 5a_{n-2} + 7$

This can be written as

$$a_n - 4a_{n-1} - 5a_{n-2} = 7 \tag{1}$$

The same can be written in quadratic form of equation as

$$x^2 - 4x^1 - 5x^0 = 0 \quad ==> \quad x^2 - 4x - 5 = 0$$

The roots for the quadratic equation are $x = 5$, and $x = -1$

So, we have two different roots and hence the $a_n(h) = u(S_1)^n + v(S_2)^n$

Here, $S_1 = 5$ and $S_2 = -1$

$$a_n(h) = u(5)^n + v(-1)^n \tag{2}$$

As, we have $a_1 = 2$ and $a_2 = 6$ in the given problem,

put $n = 1$ in (2) $==> a_1 = u(5)^1 + v(-1)^1 ==> a_1 = 5u - v$

$$2 = 5u - v \tag{3}$$

put $n = 2$ in (2) $==> a_2 = u(5)^2 + v(-1)^2 ==> a_2 = 25u + v$

$$6 = 25u + v \tag{4}$$

Solving the equations (3) and (4) simultaneously, we get

$$u = 4/15 \text{ and } v = -2/3$$

Substituting, these values of u and v in (2), we get

$$a_n(h) = 4/15 \ (5)^n - 2/3 \ (-1)^n \tag{5}$$

Recall, to find $a_n(p)$, If $f(n) = c$, then $a_n(p) = A$

So, we have  $a_n(p) = A$ $\tag{6}$

Put (6), $a_n = A$ in (1), we get   $==> A - 4A - 5A = 7$
$==> -8A = 7$

$$A = -7/8 \tag{7}$$

Put (7) in (6), we get

$$a_n(p) = -7/8 \tag{8}$$

From (5) and (8), we have

$$a_n = 4/15\ (5)^n - 2/3\ (-1)^n - 7/8$$

Hence, this is the solution for the given recurrence relation.

**Example 4:**

Solve the following recurrence relation $a_n = 6a_{n-1} - 11a_{n-2} + 6a_{n-3}$,

where $a_0 = 2$ and $a_1 = 5$ and $a_2 = 15$

**Solution:**

Recall, Linear Recurrence Relations with Constant Coefficients are of the form:

$$a_n = c_1 {}^* a_{n-1} + c_2 {}^* a_{n-2} + \ldots + c_k {}^* a_{n-k} + f(n)$$

where $c_1, c_2, \ldots, c_k$ are constants and $f(n)$ is a function of n.

So, in the given example, $f(n)$ is not present and hence $a_n = a_n(h)$

In the given problem, we have $a_n = 6a_{n-1} - 11a_{n-2} + 6a_{n-3}$

This can be written as

$$a_n - 6a_{n-1} + 11a_{n-2} - 6a_{n-3} = 0$$

The same can be written in cubic form of equation as

$$x^3 - 6x^2 + 11x^1 - 6x^0 = 0 \quad ==> \quad x^3 - 6x^2 + 11x - 6 = 0$$

The roots for the cubic equation are $x = 1$, $x = 2$ and $x = 3$

So, we have three different roots and hence

$$a_n(h) = u(S_1)^n + v(S_2)^n + w(S_3)^n$$

As we have in this case $a_n = a_n(h)$, $a_n = u(S_1)^n + v(S_2)^n + w(S_3)^n$

Here, $S_1 = 1$, $S_2 = 2$ and $S_3 = 3$

$$a_n = u(1)^n + v(2)^n + w(3)^n \qquad (1)$$

As, we have $a_0 = 2$ and $a_1 = 5$ and $a_2 = 15$ in the given problem,

put $n = 0$ in (1) $==> a_0 = u(1)^0 + v(2)^0 + w(3)^0$

$$a_0 = u + v + w$$

$$2 = u + v + w \qquad (2)$$

put $n = 1$ in (1) $==> a_1 = u(1)^1 + v(2)^1 + w(3)^1$

$$a_0 = u + 2v + 3w$$

$$5 = u + 2v + 3w \qquad (3)$$

put $n = 2$ in (1) $==> a_2 = u(1)^2 + v(2)^2 + w(3)^2$

$$a_0 = u + 4v + 9w$$

$$15 = u + 4v + 9w \qquad (4)$$

Solving the equations (2), (3) and (4) simultaneously, we get

$$u = 1 \, , \, v = -1 \text{ and } w = 2$$

Substituting, these values of u, v and w in (1), we get

$$a_n = (1)^n - (2)^n + 2(3)^n$$

Hence, this is the solution for the given recurrence relation.

**Example 5:**

Solve the following recurrence relation $a_n = -3a_{n-1} - 3a_{n-2} - a_{n-3}$,

where $a_1 = 3$ and $a_2 = 3$ and $a_3 = 7$

**Solution:**

Recall, Linear Recurrence Relations with Constant Coefficients are of the form:

$$a_n = c_1 * a_{n-1} + c_2 * a_{n-2} + \ldots + c_k * a_{n-k} + f(n)$$

where $c_1, c_2, \ldots, c_k$ are constants and $f(n)$ is a function of n.

So, in the given example, $f(n)$ is not present and hence $a_n = a_n(h)$

In the given problem, we have $a_n = -3a_{n-1} - 3a_{n-2} - a_{n-3}$

This can be written as

$$a_n + 3a_{n-1} + 3a_{n-2} + a_{n-3} = 0$$

The same can be written in cubic form of equation as

$$x^3 + 3x^2 + 3x^1 + 1x^0 = 0 \implies x^3 + 3x^2 + 3x + 1 = 0$$
$$\implies (x + 1)^3 = 0$$

The roots for the cubic equation are $x = -1$, $x = -1$ and $x = -1$

So, we have three same roots and hence

$$a_n(h) = u(S)^n + vn(S)^n + wn^2(S)^n$$

As we have in this case $a_n = a_n(h)$,     $a_n = u(S)^n + vn(S)^n + wn^2(S)^n$

Here, $S = -1$,

$$a_n = u(-1)^n + vn(-1)^n + wn^2(-1)^n \qquad (1)$$

As, we have $a_1 = 3$ and $a_2 = 3$ and $a_3 = 7$ in the given problem,

put $n = 1$ in (1) $\implies a_1 = u(-1)^1 + v(1)(-1)^1 + w(1)^2(-1)^1$

$$a_1 = -u - v - w$$

$$3 = -u - v - w \tag{2}$$

put n = 2 in (1)  ==>  $a_2 = u(-1)^2 + v(2)(-1)^2 + w(2)^2(-1)^2$

$$a_2 = u + 2v + 4w$$

$$3 = u + 2v + 4w \tag{3}$$

put n = 3 in (1)  ==>  $a_3 = u(-1)^3 + v(3)(-1)^3 + w(3)^2(-1)^3$

$$a_3 = -u - 3v - 9w$$

$$7 = -u - 3v - 9w \tag{4}$$

Solving the equations (2), (3) and (4) simultaneously, we get

$$u = 25 \, , \, v = 30 \text{ and } w = 8$$

Substituting, these values of u, v and w in (1), we get

$$a_n = 25(-1)^n + 30n(-1)^n + 8n^2(-1)^n$$

Hence, this is the solution for the given recurrence relation.

## 3.4. SUBSTITUTION METHOD

The substitution method is a technique commonly used to solve recurrence relations, which are equations that recursively define a sequence or function. The substitution method involves making an educated guess about the form of the solution and then proving it correct through mathematical induction.

Here's a step-by-step guide on how to use the substitution method to solve a recurrence relation:

**Guess the Form of the Solution:** Based on the given recurrence relation, try to guess the form of the solution. This often involves examining the structure of the recurrence relation and looking for patterns.

## Prove Correctness by Induction:

a.   **Base Case:** Start by proving that the solution holds for the base case(s) of the recurrence relation. Typically, this involves substituting the base case(s) into the guessed solution and verifying that the resulting expression is true.

b.   **Inductive Step:** Assume that the solution holds for some arbitrary value up to k. Then, using this assumption, prove that the solution also holds for k+1. This usually involves substituting k into the recurrence relation and simplifying until the solution for k+1 is obtained.

   **Conclude:** Once you've proven the base case and the inductive step, you can conclude that the guessed solution is correct for all values of n.

### Example 1:

Solve the recurrence relation using substitution method:

$$T(n) = 1 \text{ if } n = 1 \text{ and } T(n) = n * T(n-1) \text{ if } n > 1$$

### Solution:

$$\text{Given } T(n) = n * T(n-1) \tag{1}$$

$$\text{Then, } T(n-1) = (n-1) * T(n-2) \tag{2}$$

$$\text{Similarly, } T(n-2) = (n-2) * T(n-3) \tag{3}$$

Now, substituting (2) in (1), we get

$$T(n) = n * (n-1) * T(n-2) \tag{4}$$

Now, substitute (3) in (4), we get

$$T(n) = n * (n-1) * (n-2) * T(n-3)$$

Proceeding like this, the next substitution will give the pattern,

$$T(n) = n * (n-1) * (n-2) * (n-3) * T(n-4)$$

So, the $k^{th}$ step will have the pattern,

$$T(n) = n * (n-1) * (n-2) * \ldots * (n-(k-1)) * T(n-k) \text{ ----} \qquad (k)$$

But, the base case $T(1) = 1$

Hence, to arrive at the base case we need to substitute n-k = 1

That is k = n-1

Substituting this k value in (k) we get,

$$T(n) = n * (n-1) * (n-2) * \ldots * (n-(n-2)) * T(1)$$

Therefore, $T(n) = n * (n-1) * (n-2) * \ldots * 2 * 1 = n!$

Therefore the time complexity of this recurrence relation is n * n * n * … * n * n (n times) = $n^n$

That is the time complexity is $O(n^n)$.

**Example 2:**

Solve the recurrence relation using substitution method:

$$T(n) = 1 \text{ if } n = 1 \text{ and } T(n) = 2 * T(n/2) + n \text{ if } n > 1$$

**Solution:**

Given $T(n) = 2 * T(n/2) + n$ $\qquad\qquad$ (1)

Then, $T(n/2) = 2 * T(n/4) + n/2$ $\qquad\qquad$ (2)

Similarly, $T(n/4) = 2 * T(n/8) + n/4$ $\qquad\qquad$ (3)

Now, substituting (2) in (1), we get

$$T(n) = 2 * [2 * T(n/4) + n/2] + n$$

$$T(n) = 4 * T(n/4) + 2n \qquad\qquad (4)$$

Now, substitute (3) in (4), we get

$$T(n) = 4 * [2 * T(n/8) + n/4] + 2n$$

$$T(n) = 8 * T(n/8) + 3n$$

$$T(n) = 2^3 * T(n/2^3) + 3n$$

Proceeding like this, the next substitution will give the pattern,

$$T(n) = 2^4 * T(n/2^4) + 4n$$

So, the $k^{th}$ step will have the pattern,

$$T(n) = 2^k * T(n/2^k) + kn \quad ---- \qquad\qquad (k)$$

But, the base case $T(1) = 1$

Hence, to arrive at the base case we need to substitute $n/2^k = 1$

That is $n = 2^k$

Taking log on both sides, we get $\log n = \log 2^k$

That is $\log n = k \log 2$

That is $\log n = k$ (as log 2 base 2 is 1)

Substituting $2^k$ value and k value in (k) for the base case, we get

$$T(n) = n * T(1) + n \log n$$

Therefore $T(n) = n * 1 + n \log n = n + n \log n$

Therefore the time complexity of the given recurrence relation is O(n log n)

(Considering only the dominating n terms).

**Example 3:**

Solve the recurrence relation using substitution method:

$T(n) = 1$ if $n = 1$ and $T(n) = T(n-1) + \log n$ , if $n > 1$

**Solution:**

Given $T(n) = T(n-1) + \log n$     (1)

Then, $T(n-1) = T(n-2) + \log (n-1)$     (2)

Similarly, $T(n-2) = T(n-3) + \log (n-2)$     (3)

Now, substituting (2) in (1), we get

$T(n) = T(n-2) + \log (n-1) + \log n$     (4)

Now, substitute (3) in (4), we get

$T(n) = T(n-3) + \log (n-2) + \log (n-1) + \log n$

Proceeding like this, the next substitution will give the pattern,

$T(n) = T(n-4) + \log (n-3) + \log (n-2) + \log (n-1) + \log n$

So, the $k^{th}$ step will have the pattern,

$T(n) = T(n-k) + \log (n-(k-1)) + \log (n-(k-2)) + ... + \log (n-1) + \log n$ ----     (k)

But, the base case $T(1) = 1$

Hence, to arrive at the base case we need to substitute $n-k = 1$

That is $k = n-1$

Substituting this k value in (k) we get,

$$T(n) = T(n - (n-1)) + \log (n - (n-2)) + \log (n-(n-3)) + ... + \log (n-2) + \log (n-1) + \log n$$

Therefore, $T(n) = T(1) + \log (2) + \log (3) + ... + \log (n-2) + \log (n-1) + \log n$

That is, $T(n) = 1 + \log (2 * 3 * ... * (n-2) * (n-1) *n) = 1 + \log (n!)$

Therefore the time complexity of this recurrence relation is $1 + \log (n^n) = 1 + n \log n$

That is the time complexity is $O(n \log n)$.

**Example 4:**

Solve the recurrence relation using substitution method:

$$T(n) = 1 \text{ if } n = 1 \text{ and } T(n) = T(n/2) + c \text{ if } n > 1$$

**Solution:**

$$\text{Given } T(n) = T(n/2) + c \qquad (1)$$

$$\text{Then, } T(n/2) = T(n/4) + c \qquad (2)$$

$$\text{Similarly, } T(n/4) = T(n/8) + c \qquad (3)$$

Now, substituting (2) in (1), we get

$$T(n) = T(n/4) + 2c \qquad (4)$$

Now, substitute (3) in (4), we get

$$T(n) = T(n/8) + 3c$$

$$T(n) = T(n/2^3) + 3c$$

Proceeding like this, the next substitution will give the pattern,

$$T(n) = T(n/2^4) + 4c$$

So, the $k^{th}$ step will have the pattern,

$$T(n) = T(n/2^k) + kc \text{ ----} \qquad (k)$$

But, the base case $T(1) = 1$

Hence, to arrive at the base case we need to substitute $n/2^k = 1$

That is $n = 2^k$

Taking log on both sides, we get $\log n = \log 2^k$

That is $\log n = k \log 2$

That is $\log n = k$ (as log 2 base 2 is 1)

Substituting k value in (k) for the base case, we get

$$T(n) = T(1) + c \log n$$

Therefore $T(n) = 1 + c \log n$

Therefore the time complexity of the given recurrence relation is $O(\log n)$.

## 3.5. RECURRENCE TREES

Recurrence trees are graphical representations used to analyze the time complexity of recursive algorithms. When you have a recursive function, it often calls itself with smaller inputs until it reaches a base case where it returns a known value without further recursion. Recurrence trees help visualize how the function calls itself and how many times it does so for a given input size.

## Here's a basic overview of how recurrence trees work:

**Identify the Recurrence Relation:** This is the equation that describes the time complexity of the algorithm. It expresses the time taken by a problem of size 'n' in terms of the times taken by smaller subproblems.

**Draw the Recurrence Tree:** Each node in the tree represents a subproblem, and the children of a node represent the subproblems generated by dividing the original problem into smaller ones. The tree continues until it reaches the base case(s).

**Analyze the Tree:** Once you have drawn the tree, you can analyze it to determine the total time complexity of the algorithm. This usually involves counting the number of nodes at each level and summing them up.

**Solve the Recurrence Relation:** You can often express the total time complexity of the algorithm as a function of 'n' using the recurrence relation. Solving this relation gives you the time complexity of the algorithm.

**Optional:** Simplify the Tree: Sometimes, you can simplify the recurrence tree by aggregating nodes at the same level or using other techniques. This can make the analysis easier.

Recurrence trees are particularly useful for understanding the time complexity of algorithms like divide-and-conquer algorithms (e.g., merge sort, quicksort) and recursive algorithms on trees (e.g., tree traversal algorithms). They provide a visual aid that can make it easier to understand how the time complexity of a recursive algorithm evolves as the input size increases.

**Example 1:**

Consider the recurrence relation for which the time complexity has to be found using the recurrence trees method.

$$T(n) = 2T(n/2) + n \qquad\qquad (1)$$

Now the tree takes the form as below as

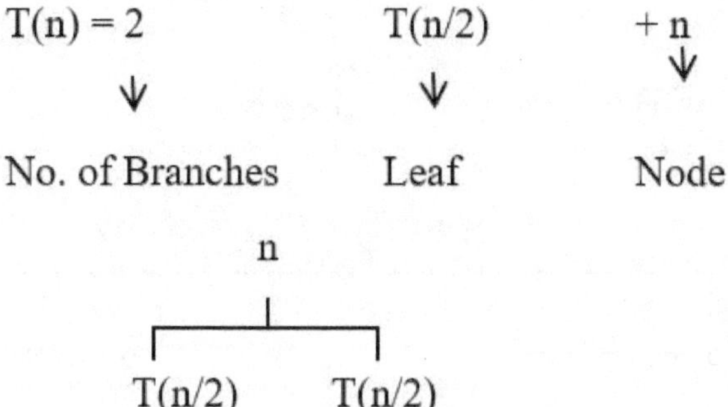

To find $T(n/2)$, we substitute $n = n/2$ in (1), so we get,

$$T(n/2) = 2T(n/4) + n/2 \qquad\qquad (2)$$

That is, $T(n/2^1) = 2\,T(n/2^2) + n/2^1$

Now the tree takes the form as below

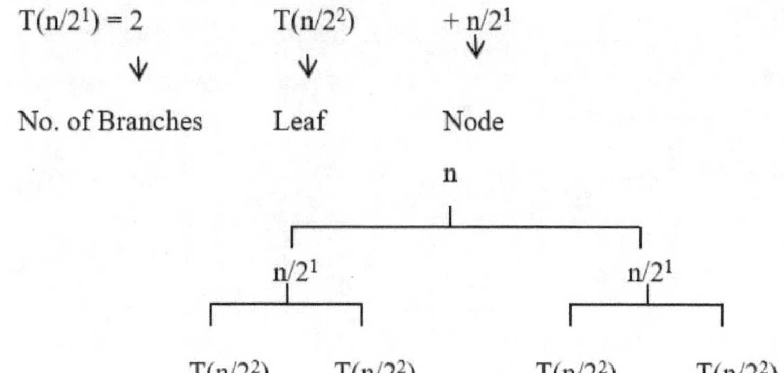

To find $T(n/4)$, we substitute $n = n/2$ in (2), so we get,

$$T(n/4) = 2T(n/8) + n/4 \qquad\qquad (3)$$

That is, $T(n/2^2) = 2\,T(n/2^3) + n/2^2$

Now the tree takes the form as below

$$T(n/2^2) = 2 \qquad T(n/2^3) \qquad + n/2^2$$

No. of Branches        Leaf        Node

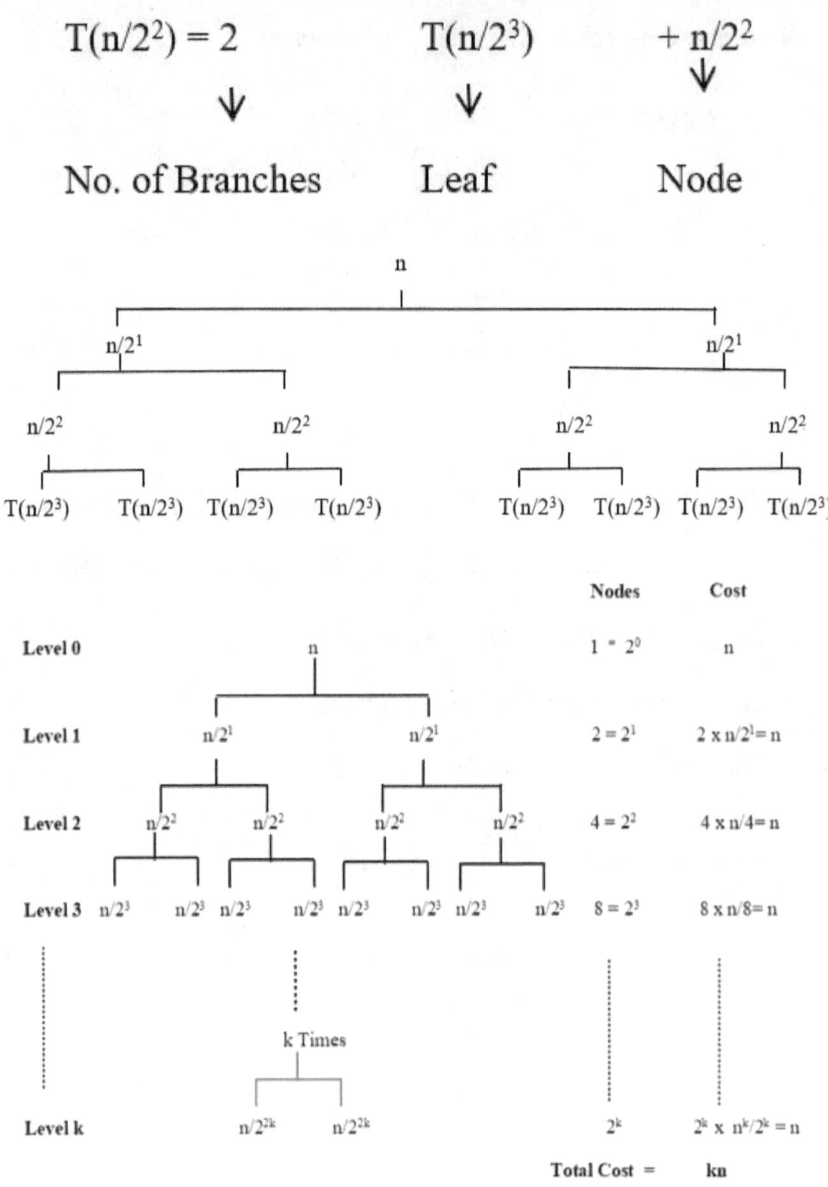

Expanding the tree like this till k times, we get the tree like below:

Therefore, the total cost is kn.

The base case is $n/2^k = 1$, $==> n = 2^k ==> \log n = \log 2^k$ $==> \log n = k \log 2 ==> k = \log n$

Therefore the time complexity of the given recurrence relation is $O(n \log n)$

**Example 2:**

Consider the recurrence relation for which the time complexity has to be found using the recurrence trees method.

$$T(n) = 2T(n/2) + n^2 \qquad (1)$$

Now the tree takes the form as below as

$$T(n) = 2 \qquad\qquad T(n/2) \qquad\qquad + n^2$$
$$\downarrow \qquad\qquad\qquad \downarrow \qquad\qquad\qquad \downarrow$$

No. of Branches          Leaf          Node

$$n^2$$

$$T(n/2) \qquad T(n/2)$$

To find $T(n/2)$, we substitute $n = n/2$ in (1), so we get,

$$T(n/2) = 2T(n/4) + n^2/4 \qquad (2)$$

That is, $T(n/2^1) = 2\, T(n/2^2) + n^2/2^2$

Now the tree takes the form as below

$T(n/2^1) = 2$      $T(n/2^2)$      $+ n^2/2^2$

No. of Branches      Leaf      Node

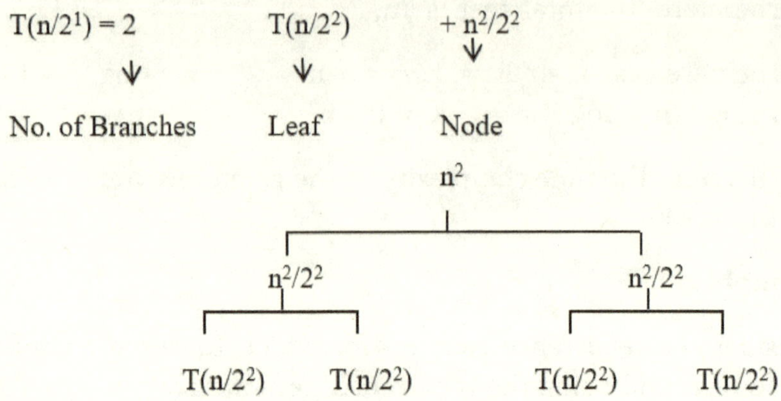

To find T(n/4), we substitute n = n/2 in (2), so we get,

$$T(n/4) = 2T(n/8) + n^2/16 \qquad (3)$$

That is, $T(n/2^2) = 2\,T(n/2^3) + n^2/2^4$

Now the tree takes the form as below

$T(n/2^2) = 2$      $T(n/2^3)$      $+ n^2/2^4$

No. of Branches      Leaf      Node

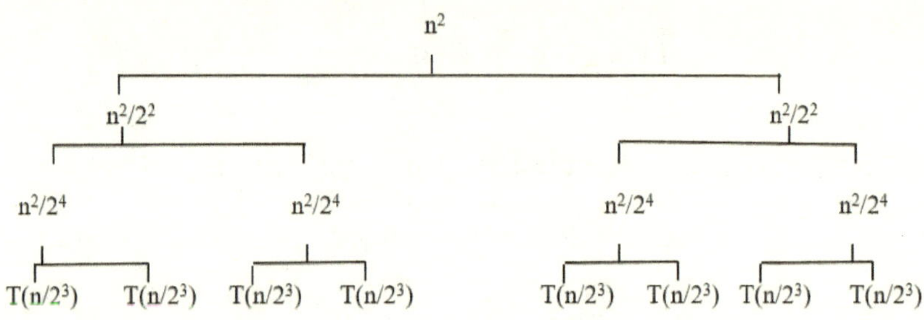

Expanding the tree like this till k times, we get the tree like below:

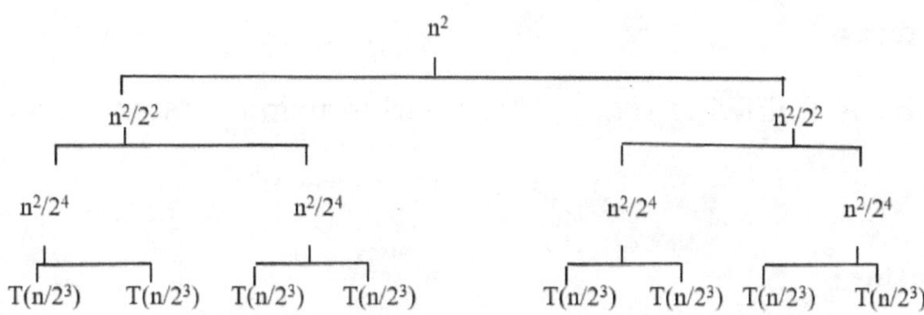


Therefore, the total cost is $2n^2$.

The base case is $n/2^k = 1$, $==> n = 2^k ==> \log n = \log 2^k$ $==> \log n = k \log 2 ==> k = \log n$

Therefore the time complexity of the given recurrence relation is $O(n^2)$

## 3.6. MASTER THEOREM

Master's theorem is a popular method for solving the recurrence relations. The recurrence relation is in the form:

$$T(n) = a * T(n/b) + \theta (n^k \log^p n)$$

Where $a >= 1, b > 1, k >= 0$ and p is a real number.

To solve the recurrence relations using Master's theorem, we compare 'a' with '$b^k$', then we follow the below cases:

**Case 1:** If $a > b^k$, then $T(n) = \theta (n^{\log_b a})$

**Case 2:** If $a = b^k$ then,

    i.    If $p < -1$, then $T(n) = \theta (n^{\log_b a})$

    ii.    If $p = -1$, then $T(n) = \theta (n^{\log_b a} * \log^2 n)$

    iii.    If $p > -1$, then $T(n) = \theta (n^{\log_b a} * \log^{p+1} n)$

**Case 3:** If $a < b^k$ then,

    i.    If $p < 0$, then $T(n) = \theta(n^k)$

    ii.    If $p >= 0$, then $T(n) = \theta(n^k * \log^p n)$

**Example 1:**

Solve the following recurrence relation using master theorem.

$$T(n) = 3 * T(n/2) + n^2$$

**Solution:**

Here $a = 3$, $b = 2$, $k = 2$, $p = 0$ and $b^k = 4$

Therefore case 3 (ii) applies as $a < b^k$ and $p >= 0$

Therefore, the time complexity for the given recurrence relation is $\theta(n^2 * \log^0 n) = \theta(n^2)$

**Example 2:**

Solve the following recurrence relation using master theorem. (Merge sort)

$$T(n) = 2 * T(n/2) + n$$

**Solution:**

Here $a = 2$, $b = 2$, $k = 1$, $p = 0$ and $b^k = 2$

Therefore case 2 (iii) applies as $a = b^k$ and $p >= -1$

Therefore, the time complexity for the given recurrence relation is $\theta(n^{\log_2 2} * \log^{0+1} n)$

That is the time complexity for the given recurrence relation is $\theta(n^1 * \log n) = \theta(n \log n)$

## Example 3:

Solve the following recurrence relation using master theorem. (Merge sort)

$$T(n) = 2 * T(n/2) + n \log n$$

**Solution:**

Here a = 2, b = 2, k = 1, p = 1 and $b^k$ = 2

Therefore case 2 (iii) applies as a = $b^k$ and p >= -1

Therefore, the time complexity for the given recurrence relation is $\theta\,(n^{\log_2 2} * \log^{1+1} n)$

That is the time complexity is $\theta\,(n^1 * \log^2 n) = \theta\,(n \log^2 n)$

## Example 4:

Solve the following recurrence relation using master theorem.

$$T(n) = \sqrt{2} * T(n/2) + \log n$$

**Solution:**

Here a = $\sqrt{2}$, b = 2, k = 0, p = 1 and $b^k$ = 1

Therefore case 1 applies as a > $b^k$

Therefore, the time complexity for the given recurrence relation is $\theta\,(n^{\log_2 \sqrt{2}})$

That is the time complexity is $\theta\,(n^{\log_2 2^{1/2}}) = \theta\,(n^{1/2 \log_2 2}) = \theta\,(n^{1/2})$ $\theta = (\sqrt{n})$

## Example 5 :

Solve the following recurrence relation for Binary search using master theorem.

$$T(n) = T(n/2) + k$$

## Solution:

Here a = 1, b = 2, k = 0, p = 0 and $b^k$ = 1

Therefore case 2 (iii) applies as a = $b^k$ and p > -1

Therefore, the time complexity for the given recurrence relation is $\theta$ (n $^{\log_2 1}$ * log $^{0+1}$ n) = $\theta$ (n$^0$ * log n) = $\theta$ (log n)

# Chapter 4

# GRAPH THEORY

Graph theory is a branch of mathematics that deals with the study of graphs, which are mathematical structures representing relationships between a collection of points, called vertices, and the connections between them, known as edges. In graph theory, vertices and edges are used to model and analyze various real-world scenarios and abstract systems. The field encompasses a wide range of topics, including the analysis of network structures, connectivity, paths, cycles, and other properties inherent to graphs. Graph theory has applications in diverse fields such as computer science, biology, social sciences, transportation, and optimization problems, making it a fundamental and widely used area of mathematical study.

## 4.1. BASIC TERMINOLOGY

**Point**

A point denotes a specific location within one, two, or three dimensions, often symbolized by a letter for clarity. This location can be visualized as a dot.

**Example**

• a

In this instance, the dot represents a point designated as 'a'.

## Line

A line is formed by linking two points together, typically depicted as a solid line.

**Example**

In this context, 'a' and 'b' represent the points. The connection between them is referred to as a line.

## Vertex

A vertex serves as the meeting point for multiple lines and is also known as a node. Like points, vertices are typically identified using alphabetic notation.

**Example**

In this instance, the vertex is designated by the letter 'a'.

## Edge

An edge refers to the mathematical concept of a line connecting two vertices. Several edges can originate from a single vertex. Without a vertex, an edge cannot exist; it requires both a starting and an ending vertex to be defined.

**Example**

## Graph

A graph 'G' is represented as G = (V, E), where V denotes the set of all vertices, and E represents the set of all edges within the graph.

## Example1

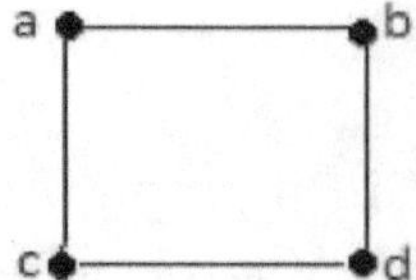

In the preceding example, the edges of the graph are represented by ab, ac, cd, and bd. Likewise, a, b, c, and d denote the vertices of the graph.

## Example 2

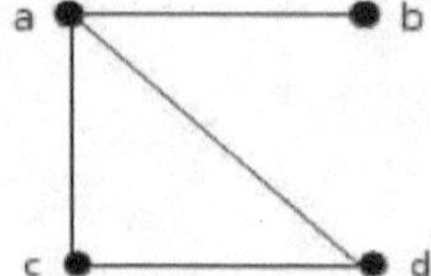

Within this graph, there exist four vertices: a, b, c, and d, accompanied by four edges: ab, ac, ad, and cd.

## Loop

Within a graph, when an edge is drawn from a vertex back to itself, it is termed a loop.

## Example 1

In the graph shown above, vertex V possesses an edge (V, V), creating a loop.

**Example 2**

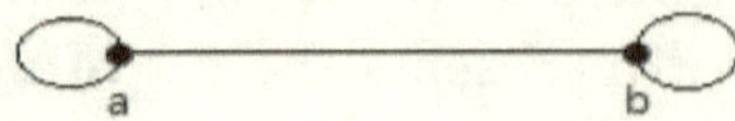

Within this graph, two loops are formed at vertices a and b.

**Degree of Vertex**

The degree of a vertex V in a graph, denoted as deg(V), is such that deg(V) $\leq$ n - 1 for all V $\in$ G, where n represents the total number of vertices in the graph.

## 4.2. MODELS AND TYPES

While there exists a multitude of graph variations determined by factors such as the number of vertices, edges, interconnections, and overall structure, several typical types can be identified, including:

1) **Null Graph**

   A null graph, also known as an empty graph, is characterized by the absence of edges between its vertices.

   **Example**

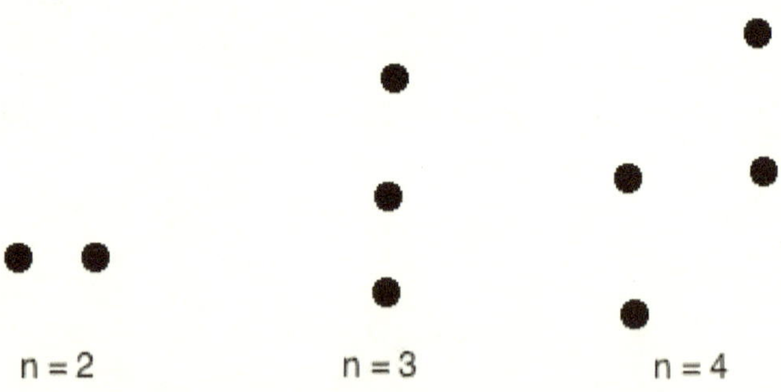


   A null graph with n vertices is denoted by $N_n$.

## 2)  Trivial Graph

A trivial graph consists of only a single vertex.

**Example**

•V

In the depicted graph, the sole vertex 'v' exists without any edges connected to it. Consequently, it qualifies as a trivial graph.

## 3)  Simple Graph

A simple graph is an undirected graph devoid of parallel edges or loops. In a simple graph with n vertices, the degree of each vertex is at most n - 1.

**Example**

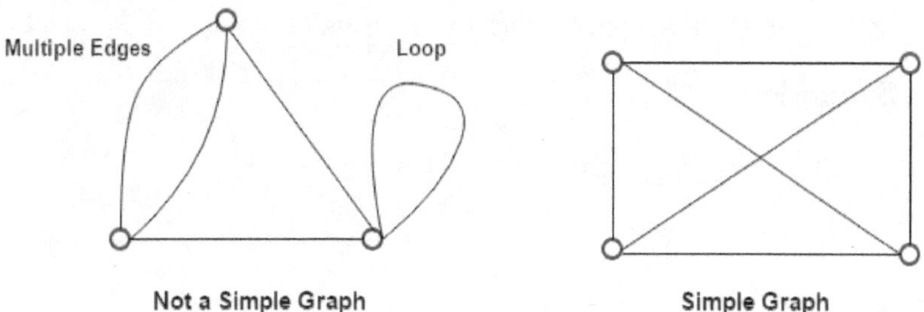


In the provided example, the first graph does not qualify as a simple graph due to the presence of two edges between vertices A and B, as well as a loop. Conversely, the second graph meets the criteria of a simple graph by lacking any loops or parallel edges.

## 4)  Undirected Graph

An undirected graph is characterized by edges that lack directionality.

**Example**

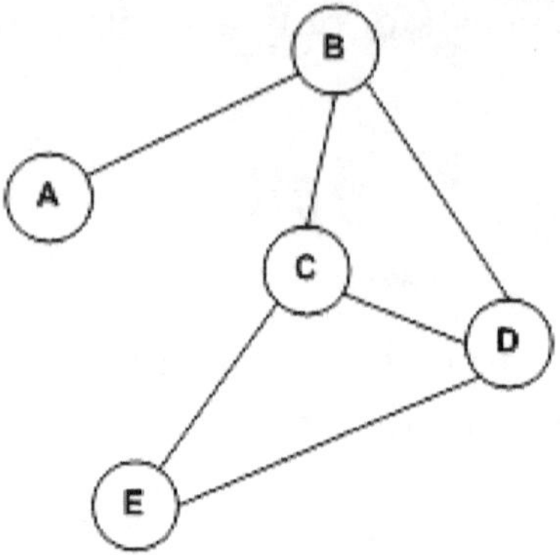

In the depicted graph, the absence of directed edges indicates that it is an undirected graph.

## 5)  Directed Graph

A directed graph, also referred to as a digraph, is distinguished by edges that are directed by arrows.

**Example**

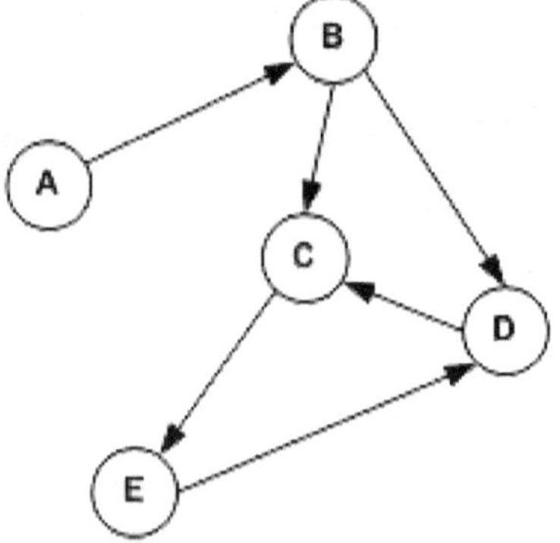

In the depicted graph, each edge is indicated by an arrow, implying direction. A directed edge with an arrow from A to B signifies a relationship from A to B, but not from B to A.

## 6)   Complete Graph

A graph where every pair of vertices is connected by precisely one edge is termed a complete graph. It encompasses all conceivable edges. A complete graph with n vertices contains exactly $nC_2$ edges and is represented by $K_n$.

**Example**

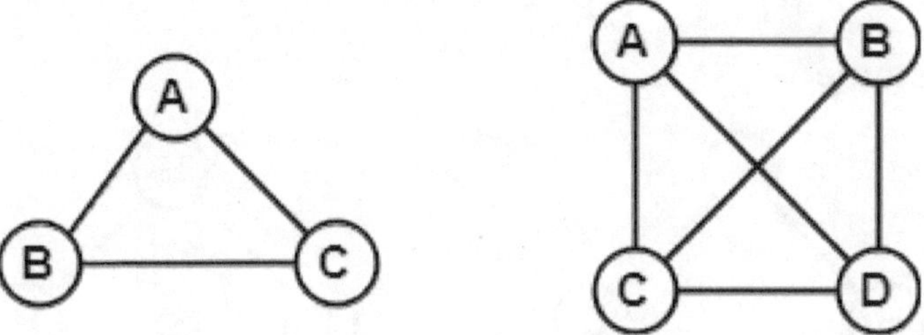

In the provided example, because every vertex in the graph is linked to all other vertices by exactly one edge, both graphs qualify as complete graphs.

## 7)   Connected Graph

A connected graph is one where it's possible to navigate from any vertex to any other vertex. In such a graph, there exists at least one edge or path connecting every pair of vertices.

**Example**

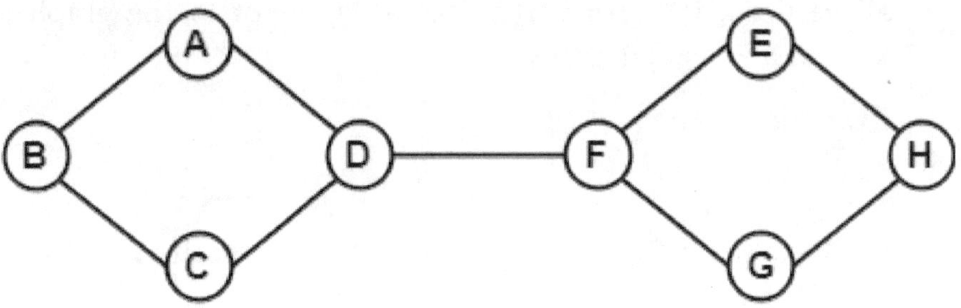

In the depicted example, it's possible to move from any vertex to any other vertex, indicating the presence of at least one path between every pair of vertices. Hence, it qualifies as a connected graph.

## 8)  Disconnected Graph

A disconnected graph is characterized by the absence of a path between every pair of vertices.

**Example**

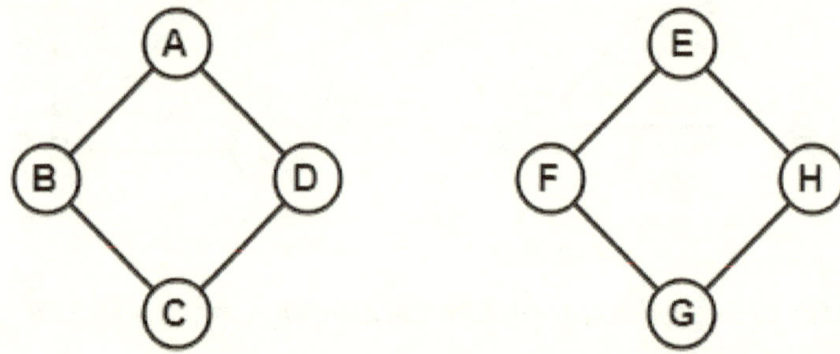

The depicted graph comprises two separate components that are disconnected. As it's not feasible to travel from the vertices of one component to those of the other, it qualifies as a disconnected graph.

## 9)  Regular Graph

A regular graph is defined by having the same degree for all vertices. If every vertex has a degree of k, the graph is termed a k-regular graph.

**Example**

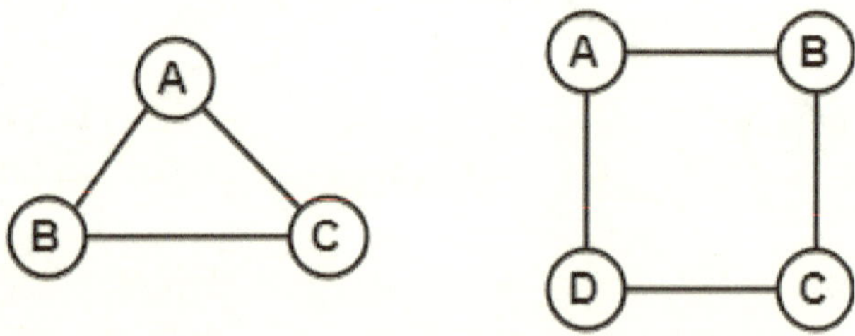

In the provided example, each vertex has a degree of 2, thus classifying it as a 2-regular graph.

## 10) Cyclic Graph

A graph with 'n' vertices (where n≥3) and 'n' edges, forming a cycle of length 'n', is referred to as a cycle graph. A graph that includes at least one cycle is termed a cyclic graph. In a cycle graph, each vertex has a degree of 2. The cycle graph which has n vertices is denoted by $C_n$.

**Example 1**

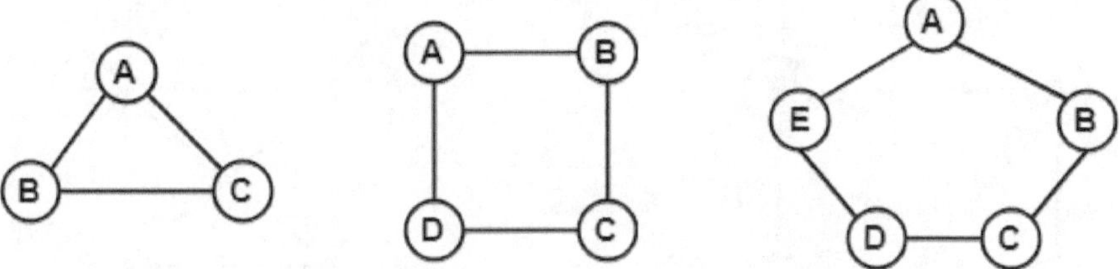

In the depicted example, every vertex possesses a degree of 2, indicating that they all represent cyclic graphs.

## 11) Acyclic Graph

An acyclic graph is defined as a graph that lacks any cycles.

**Example**

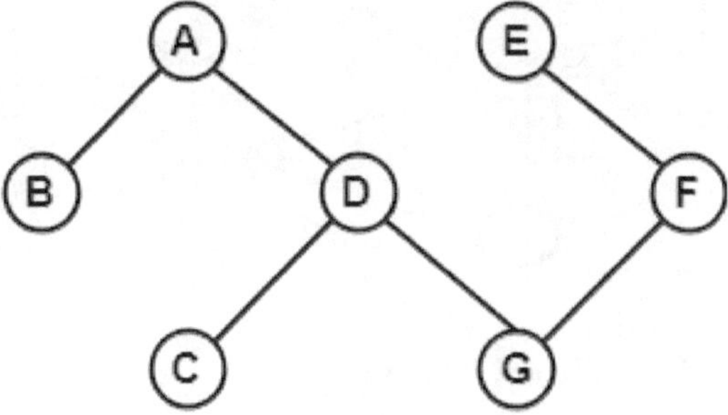

Because the depicted graph lacks any cycles, it is classified as an acyclic graph.

## 12)  Bipartite Graph

A bipartite graph is characterized by its vertex set being partitioned into two sets, where edges only connect vertices between these sets, not within them. Formally, a graph G (V, E) is deemed bipartite if its vertex set V(G) can be divided into two non-empty disjoint subsets, V1(G) and V2(G), in a manner that each edge e $\in$ E(G) connects a vertex from V1(G) to a vertex from V2(G). This partition, denoted as V = V1 $\cup$ V2, is referred to as the bipartition of G.

**Example 1**

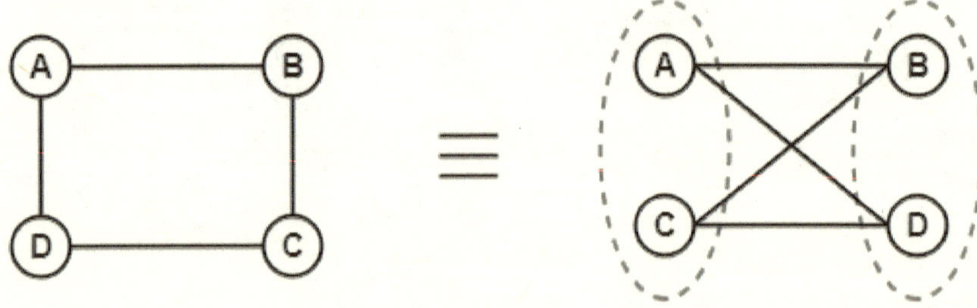

**Example 2**

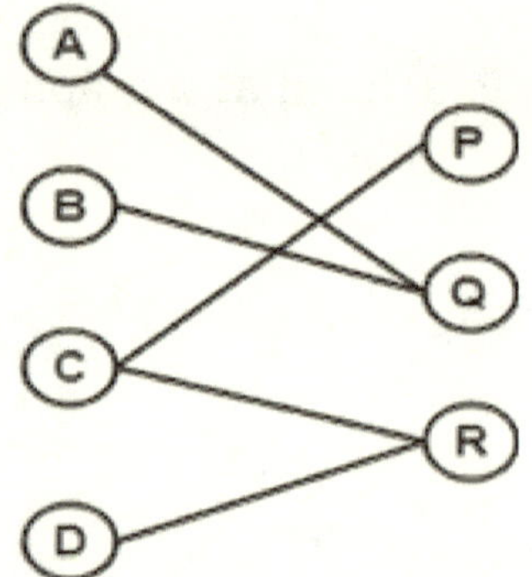

## 13)  Complete Bipartite Graph

A complete bipartite graph is characterized by every vertex in the first set being connected to each vertex in the second set by precisely one edge. It represents a bipartite graph that is also complete. In essence, a complete bipartite graph combines the properties of a bipartite graph and a complete graph.

**Example**

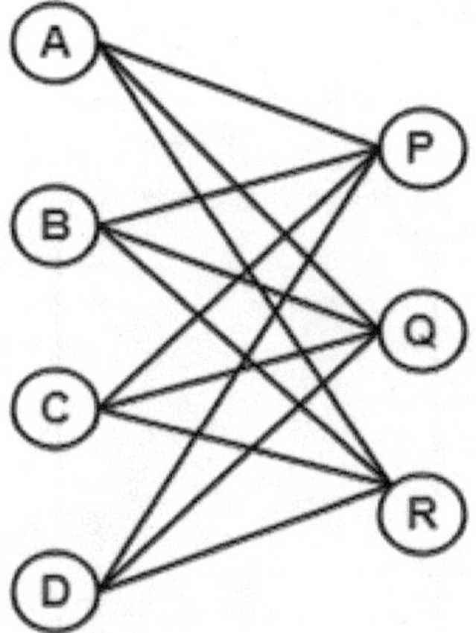

The above graph is known as $K_{4,3}$.

## 14) Star Graph

A star graph is essentially a complete bipartite graph where (n - 1) vertices have a degree of 1, and a single vertex has a degree of (n - 1). It resembles a star formation, with (n - 1) vertices radiating out from a central vertex. A star graph with n vertices is denoted by $S_n$.

**Example**

In the depicted example, all but one of the n vertices are linked to a single vertex. Therefore, it conforms to the structure of a star graph.

## 15) Weighted Graph

A weighted graph is characterized by edges that are assigned numerical values or weights. In such a graph, the length of a path is determined by the cumulative sum of the weights of all the edges within that path.

**Example**

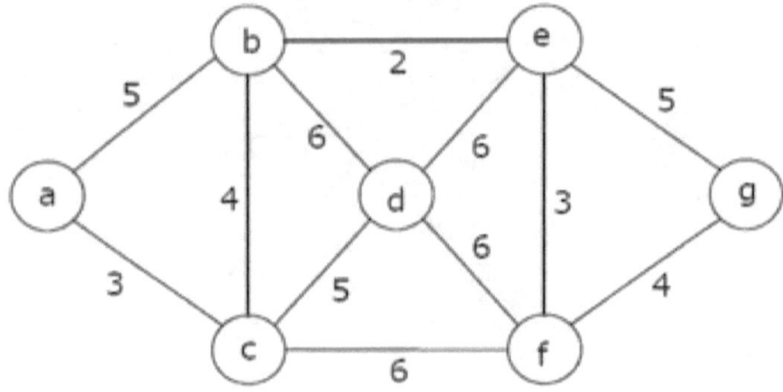

In the above graph, if path is a -> b -> c -> d -> e -> g then the length of the path is 5 + 4 + 5 + 6 + 5 = 25.

## 16) Multi-graph

A multigraph is defined as a graph containing multiple edges between any pair of vertices or edges connecting a vertex to itself (loop).

**Example**

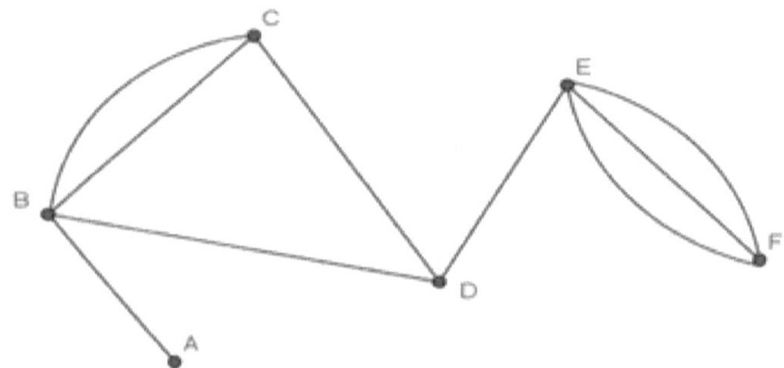

In the depicted graph, the vertex sets B and C are linked by two edges, while the vertex sets E and F are connected by three edges. Consequently, it qualifies as a multigraph.

## 17) Planar Graph

A planar graph is a graph that can be represented in a plane in a manner where edges do not intersect each other except at the vertices to which they are directly connected.

**Example**

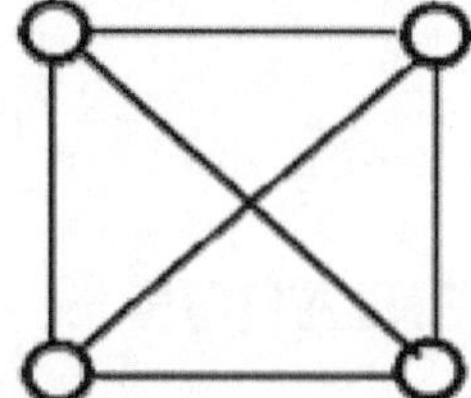

The depicted graph might not initially appear planar due to intersecting edges. However, by redrawing it, we can achieve three planar representations as follows:

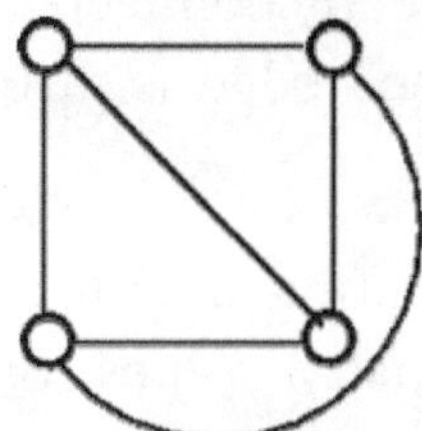
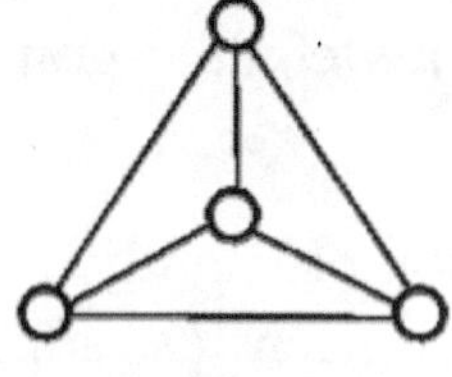
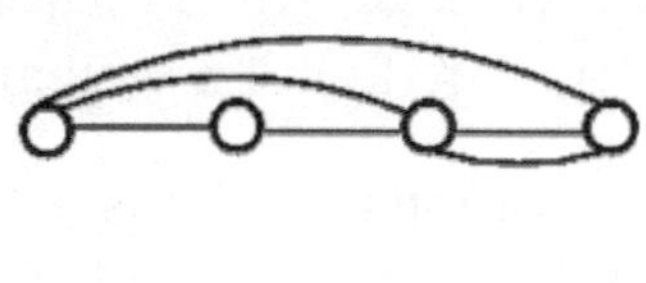

The three graphs depicted above do not contain any crossing edges, indicating that they are all planar.

## 18) Non - Planar Graph

A graph that cannot be represented as a planar graph is termed a non-planar graph. In simpler terms, a graph that unavoidably has at least one pair of crossing edges when drawn is classified as non-planar.

**Example**

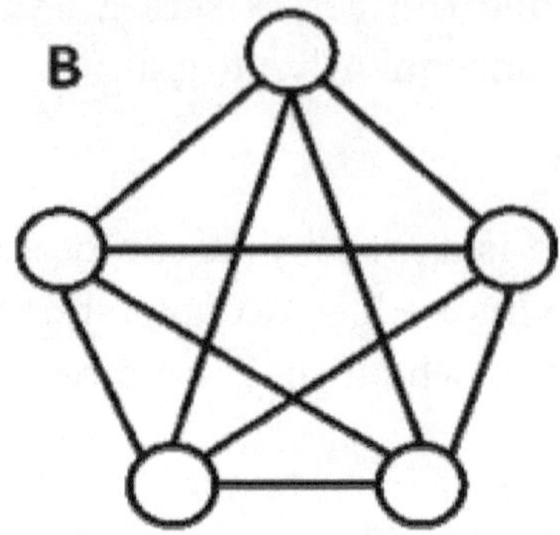



The above graph is a non - planar graph.

## 4.3. GRAPH REPRESENTATION

In the realm of graph theory, a graph representation entails a method for storing a graph in a computer's memory.

To represent a graph, we only require the set of vertices and, for each vertex, its neighboring vertices (those directly connected by an edge). In the case of a weighted graph, each edge is also associated with a weight.

Various methods exist for efficiently representing a graph, taking into account factors such as edge density, types of operations to be executed, and user-friendliness.

1)  **Adjacency Matrix**

- The adjacency matrix serves as a linear representation.

- It indicates the adjacency relationships between nodes, essentially denoting whether there are any edges connecting nodes within a graph.

- In this representation, we have to construct a nXn matrix A. If there is any edge from a vertex i to vertex j, then the corresponding element of A, $a^{i,j} = 1$, otherwise $a^{i,j} = 0$.

- For a weighted graph, rather than storing 1s and 0s, we can record the weight of each edge.

**Example**

Consider the following **undirected graph representation:**

**Undirected graph representation**

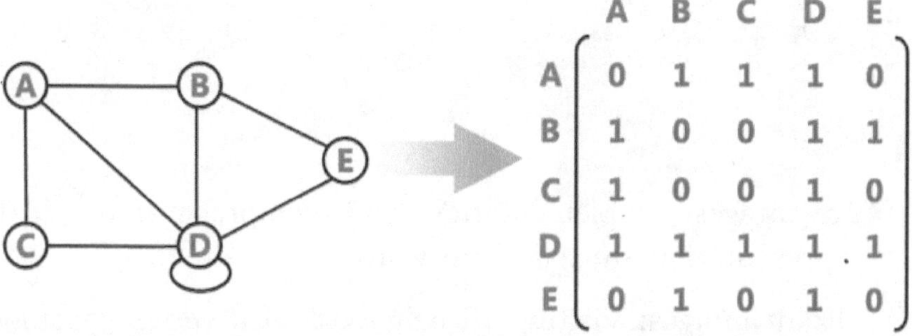

**Directed graph representation**

See the directed graph representation:

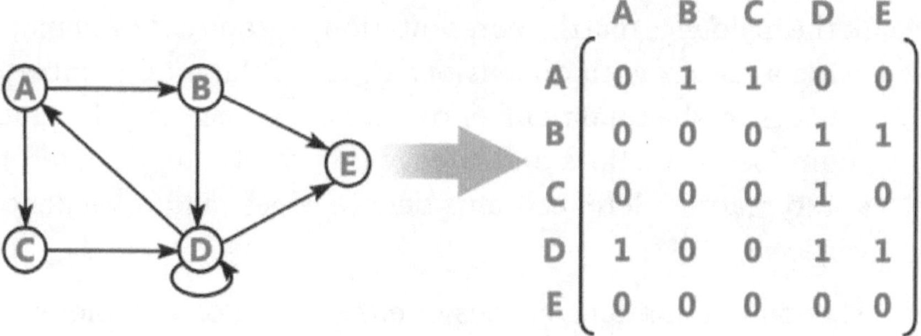

In the provided examples, a value of 1 indicates the presence of an edge from the vertex corresponding to the row to the vertex corresponding to the column, while a value of 0 indicates the absence of such an edge.

## Undirected weighted graph representation

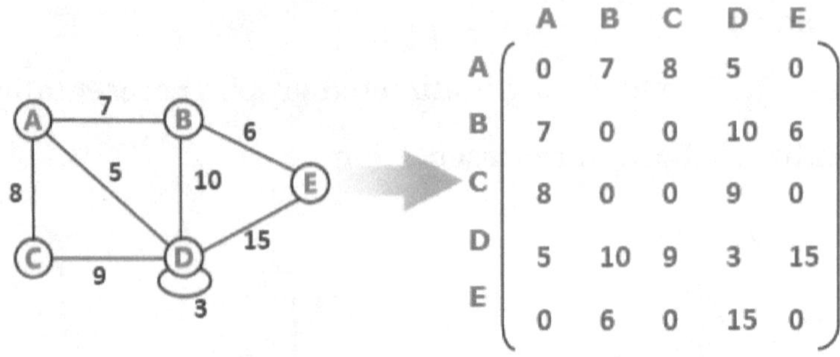

Advantages: Implementation and comprehension of the representation are straightforward.

Disadvantages: Visiting all neighbors of a vertex consumes significant space and time as it necessitates traversing all vertices in the graph, resulting in considerable time consumption.

## 2)   Incidence Matrix

In the incidence matrix representation, a graph can be depicted using a matrix with dimensions equal to the total number of vertices by the total number of edges. For instance, if a graph comprises 4 vertices and 6 edges, it can be represented by a 4x6 matrix. Here, columns denote edges, and rows denote vertices.

This matrix contains entries of either 0, 1, or -1, where:

- 0 signifies that the row vertex is not linked to the column edge.

- 1 signifies that the row vertex is connected as an outgoing edge to the column edge.

- -1 signifies that the row vertex is connected as an incoming edge to the column edge.

**Example**

Consider the following directed graph representation.

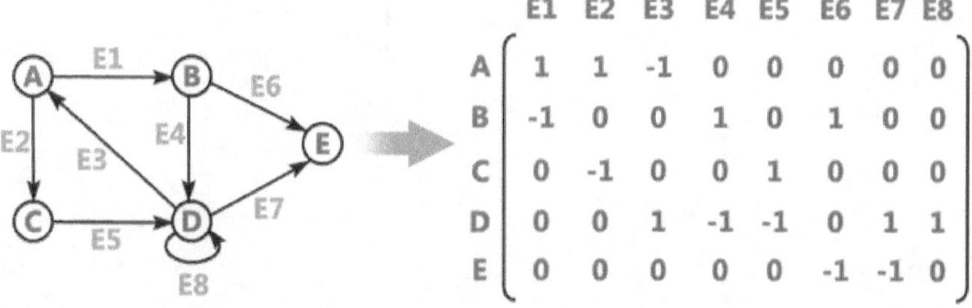

## 3)   Adjacency List

- An adjacency list is a form of linked representation.

- In this method, for every vertex in the graph, we store a list of its neighboring vertices. Essentially, each vertex in the graph maintains a list of its adjacent vertices.

- We employ an array of vertices indexed by their respective vertex numbers, where each vertex v points to a singly linked list containing its neighboring vertices.

**Example**

Let's see the following directed graph representation implemented using linked list:

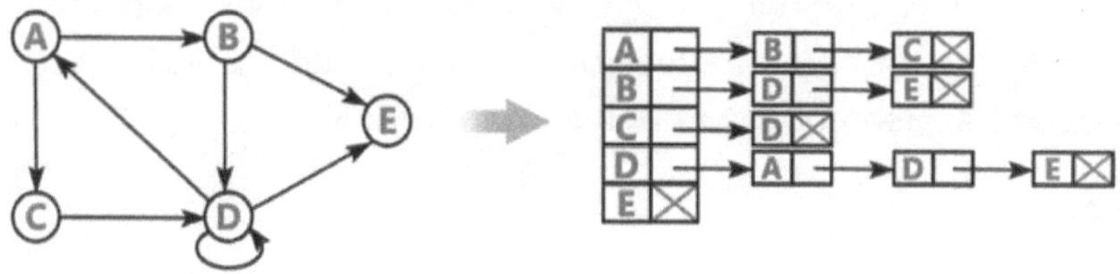

We can also implement this representation using array as follows:

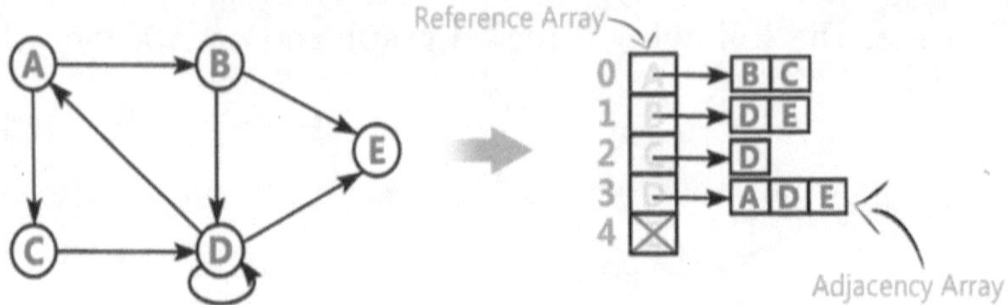

Advantages:

- Adjacency lists are space-efficient.

- They facilitate easy insertion and deletion due to the use of linked lists.

- This representation is straightforward and provides clear visibility of a node's adjacent nodes.

Disadvantages:

- Although adjacency lists enable testing for adjacency between vertices, this operation tends to be slower compared to other representations.

## 4.4. GRAPH ISOMORPHISM

Two graphs G = (V, E) and G' = (V', E') are said to be isomorphic if there exists one to one correspondence f from V to V' such that if V1, V2, $\in$ V and V1', V2' $\in$ V' and f(V1) = V1' and f(V2') = V2' then the number of edges between V1' and V2'. The function f is called an isomorphism between G and G'.

**Note:**

Two graphs are isomorphic if :

1. They must have same number of vertices

2. They must have same number of edges

3.  They must have the same degree of vertices

4.  Adjacency is preserved

**Example 1**

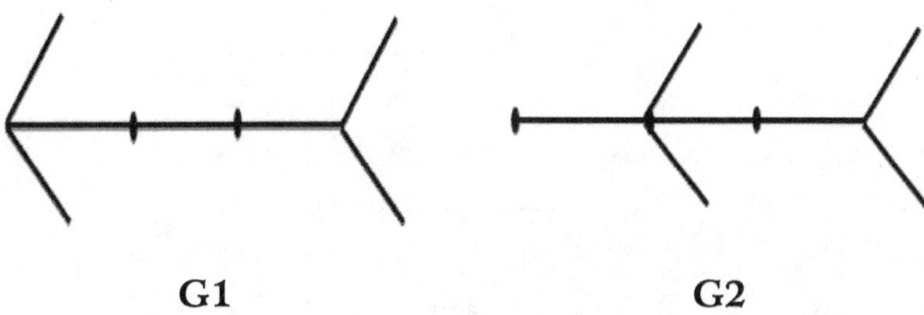

**G1**                    **G2**

In the above example 1, G1 and G2 has 8 vertices and 7 edges each. So, condition 1 and 2 are satisfied. In graph G1 there are 4 vertices with degree 1 and in graph G2 there are 5 vertices with degree 1. So, condition 3 not satisfied. Hence G1 and G2 are not isomorphic.

**Example 2**

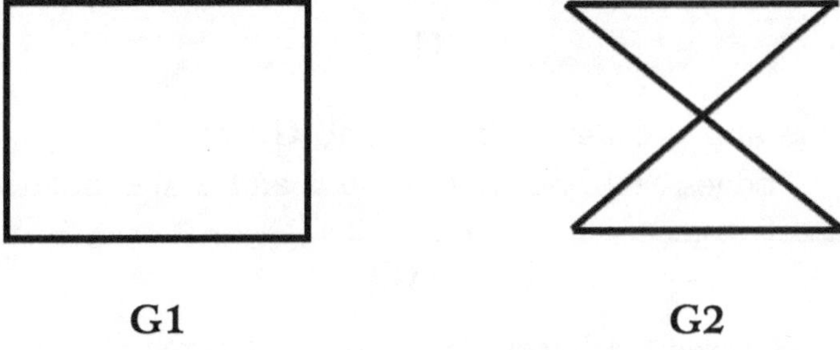

**G1**                    **G2**

In the above example 2, G1 and G2 are having 4 vertices and 4 edges each. So, Condition 1 and 2 are satisfied. In graph G1 and G2 there are 4 vertices with degree 2. So, condition 3 is also satisfied. Adjacency is also preserved in G1 and G2 as all the 4 vertices in the two graphs have degree 2. Hence G1 and G2 are isomorphic as all four conditions are satisfied.

**Example 3**

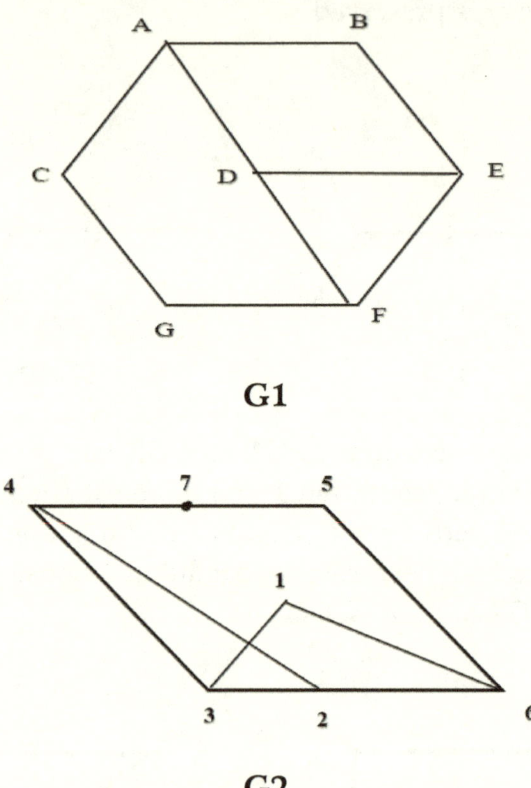

In the above example 3, G1 and G2 are having 7 vertices and 9 edges each. So, Condition 1 and 2 are satisfied.

| G1 | | | G2 | | |
|---|---|---|---|---|---|
| Vertex | Degree | Adjacent Vertices Degree | Vertex | Degree | Adjacent Vertices Degree |
| A | 3 | 2, 2, 3 | 1 | 2 | 3, 3 |
| B | 2 | 3, 3 | 2 | 3 | 3, 3, 3 |
| C | 2 | 3, 2 | 3 | 3 | 3, 3, 2 |

| G1 | | | G2 | | |
| --- | --- | --- | --- | --- | --- |
| Vertex | Degree | Adjacent Vertices Degree | Vertex | Degree | Adjacent Vertices Degree |
| D | 3 | 3, 3, 3 | 4 | 3 | 3, 3, 2 |
| E | 3 | 3, 3, 2 | 5 | 2 | 3, 2 |
| F | 3 | 3, 3, 2 | 6 | 3 | 3, 2, 2 |
| G | 2 | 3, 2 | 7 | 2 | 3, 2 |

From the above table we can say that the degrees of vertices of G1 and G2 are same (Number of degree 2 vertices is 3 and number of degree 3 vertices is 4) and hence the condition 3 is satisfied. Also, the condition 4 is satisfied as the adjacent vertices match for the below pair of vertices in G1 and G2. (A = 6; B = 1; C = 5; D = 2; E = 4; F = 3; G = 7). So, graphs G1 and G2 are isomorphic.

**Example 4**

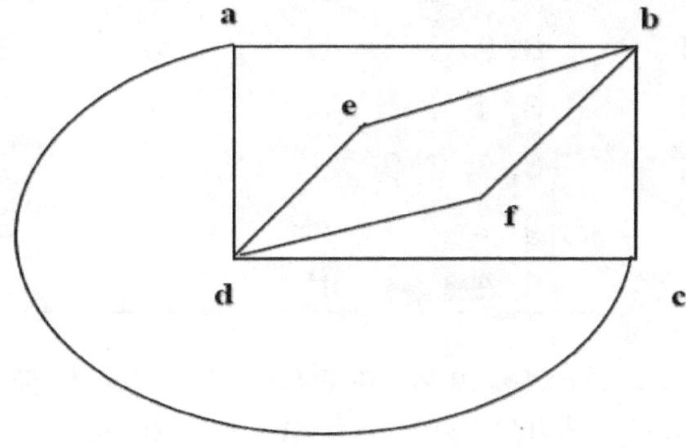


**G1**

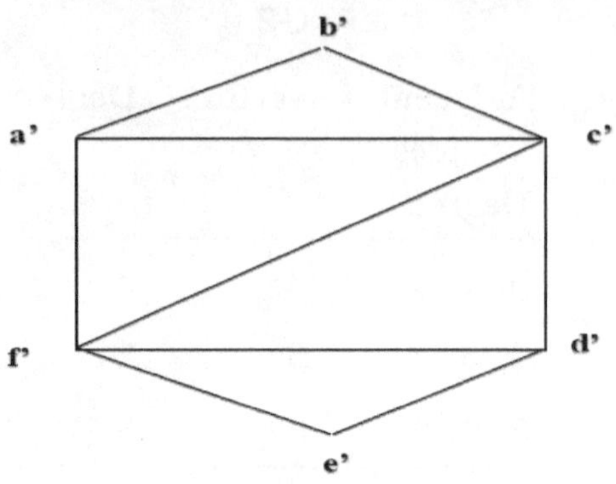

**G2**

In the above example 4, G1 and G2 are having 6 vertices and 9 edges each. So, Condition 1 and 2 are satisfied.

| G1 | | | G2 | | |
|---|---|---|---|---|---|
| Vertex | Degree | Adjacent Vertices Degree | Vertex | Degree | Adjacent Vertices Degree |
| a | 3 | 3, 4, 4 | a' | 3 | 2, 4, 4 |
| b | 4 | 3, 2, 2, 3 | b' | 2 | 3, 4 |
| c | 3 | 3, 4, 4 | c' | 4 | 2, 3, 4, 3 |
| d | 4 | 3, 2, 2, 3 | d' | 3 | 2, 4, 4 |
| e | 2 | 4, 4 | e' | 2 | 3, 4 |
| f | 2 | 4, 4 | f' | 4 | 3, 4, 2, 3 |

From the above table we can say that the degrees of vertices of G1 and G2 are same (Number of degree 2 vertices is 2, number of degree 3 vertices is 2 and number of degree 4 vertices is 2) and hence the condition 3 is satisfied. But, the condition 4 is not satisfied as the adjacent vertices degrees does not match for the graphs G1 and G2. Hence graphs G1 and G2 are not isomorphic.

## 4.5. CONNECTIVITY

Connectivity forms a fundamental aspect of graph theory, delineating whether a graph is connected or disjointed. Without connectivity, traversing from one vertex to another within a graph is unfeasible.

A connected graph is one where a path exists between every pair of vertices. From any vertex, there must be at least one path leading to any other vertex, exemplifying the graph's connectivity.

Conversely, a graph is considered disconnected if it comprises multiple disconnected vertices and edges.

The theories of graph connectivity play pivotal roles in various applications such as network routing, transportation networks, and network resilience.

**Example**

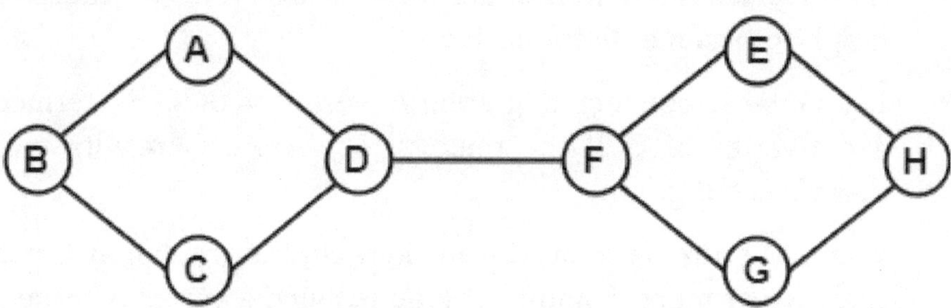


In the depicted example, navigating from one vertex to another is achievable. For instance, a route from vertex B to vertex H can be traced through the path B -> A -> D -> F -> E -> H. Consequently, the graph is classified as connected.

**Example**

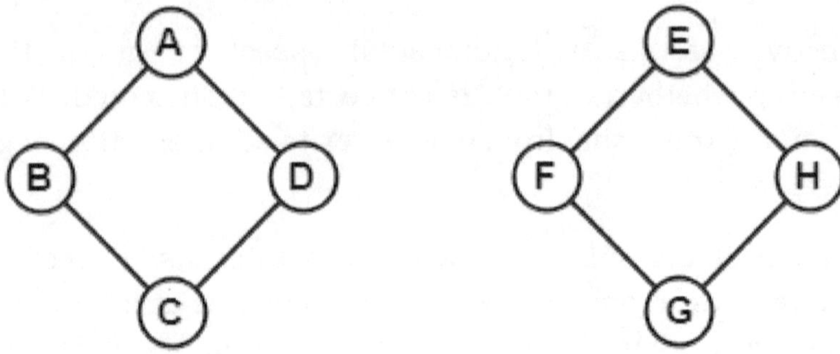

In the given example, navigating from vertex B to vertex H is not feasible as there is no direct or indirect path connecting them. Therefore, the graph is categorized as disconnected.

Let's explore some fundamental concepts of connectivity.

### 1)  Cut Vertex

A cut-vertex refers to a single vertex whose removal causes a graph to become disconnected.

Let G be a connected graph. A vertex v of G is termed a cut-vertex of G if the removal of v from G results in a disconnected graph.

When a vertex is removed from a graph, the graph may break into two or more disjoint subgraphs. Such a vertex is termed a cut-vertex.

Note: Consider a graph G with n vertices:

- A connected graph G can have a maximum of (n-2) cut vertices.

- Eliminating a cut vertex may result in a disconnected graph.

- Removing a vertex may augment the number of components in a graph by at least one.

- Each non-pendant vertex of a tree is regarded as a cut vertex.

**Example 1**

Original graph:

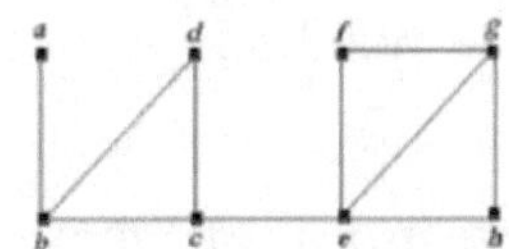

Vertex c is a cut vertex:

Vertex b is a cut vertex:

Vertex e is a cut vertex:

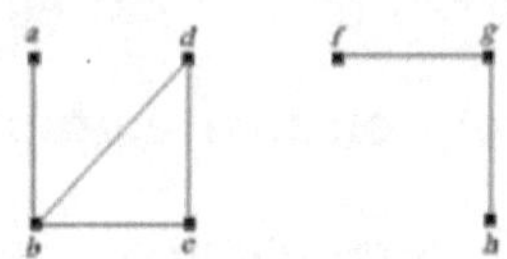

**Example 2**

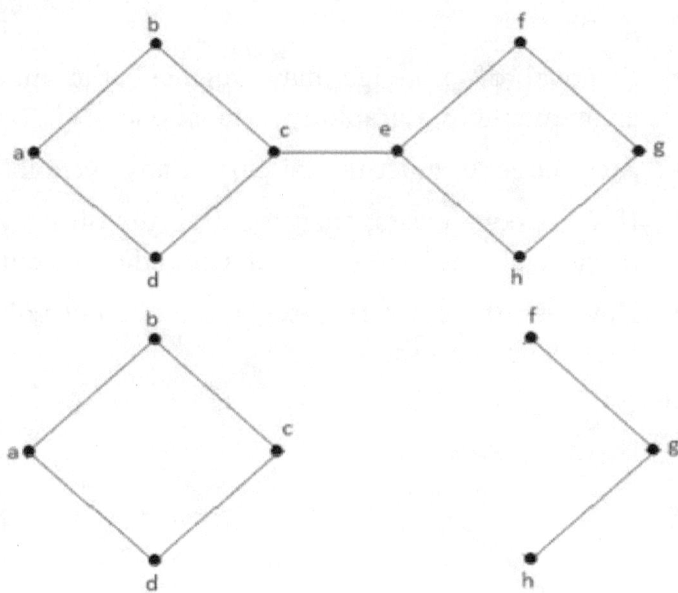

- In the depicted graph, the vertex 'e' serves as a cut vertex. Upon removal of vertex 'e' from the graph, it will become disconnected.

## 2)  Cut Edge (Bridge)

A cut edge, also known as a bridge, refers to a single edge whose removal causes a graph to become disconnected.

Let G be a connected graph. An edge e of G is designated as a cut edge of G if removing e from G results in a disconnected graph.

Upon removing an edge from a graph, it may fragment into two or more disjoint subgraphs. This removed edge is referred to as a cut edge or bridge.

Note: Consider a graph G with n vertices:

- A connected graph G can have a maximum of (n-1) cut edges.

- Eliminating a cut edge may lead to a disconnected graph.

- Removal of an edge may augment the number of components in a graph by at most one.

- A cut edge 'e' must not be part of any cycle in G.

- If a cut edge exists, then a cut vertex must also exist because at least one vertex of a cut edge is a cut vertex.

- However, if a cut vertex exists, the existence of any cut edge is not necessary.

**Example 1**

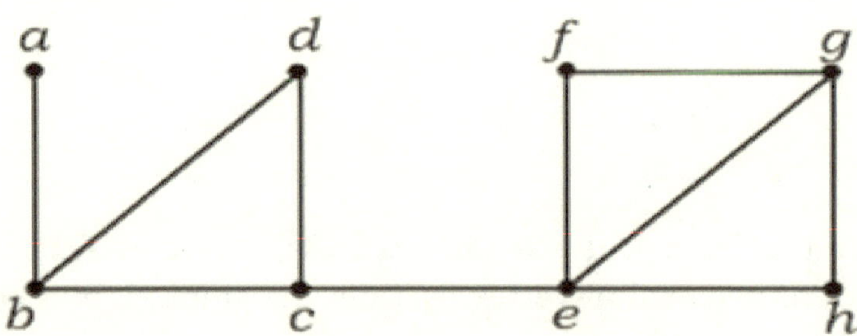

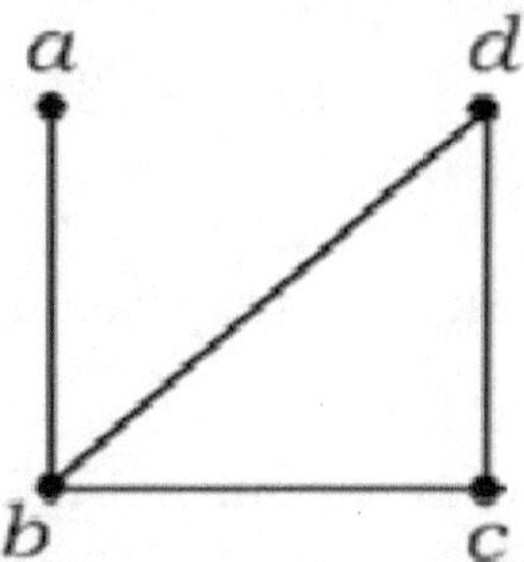 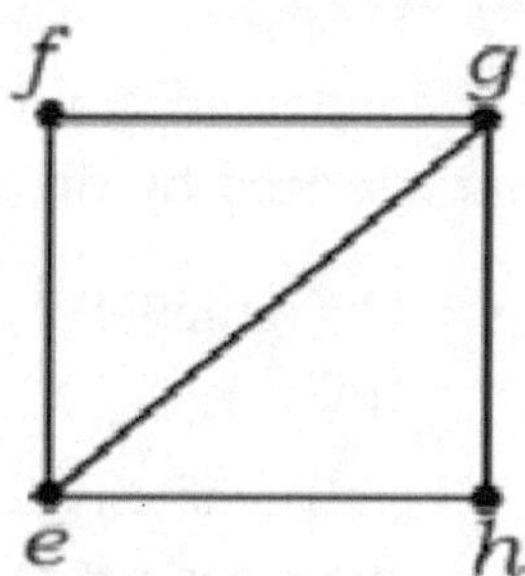

In the depicted graph, the edge (c, e) serves as a cut edge. Upon removing this edge from the graph, it will become disconnected.

**Example 2**

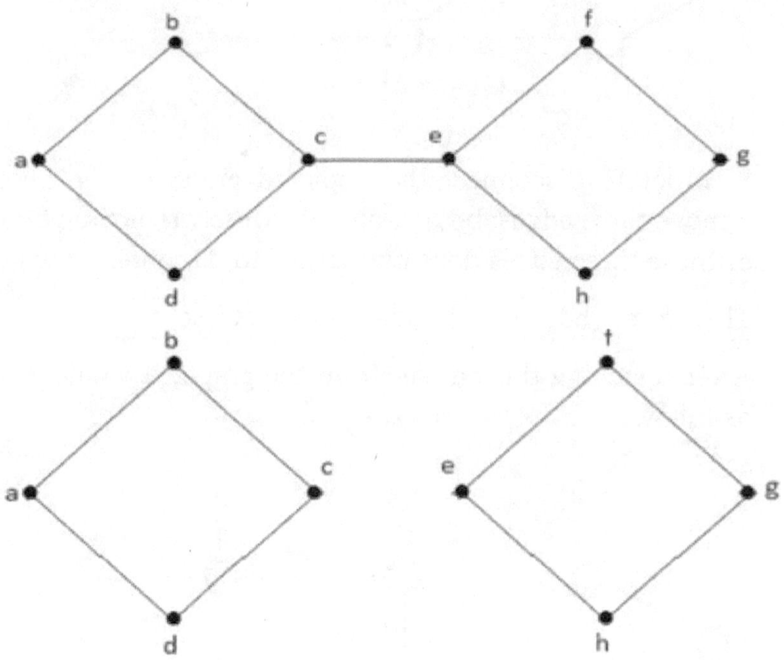

In the depicted graph, the edge (c, e) acts as a cut edge. Upon its removal, the graph will become disconnected.

## 3)   Cut Set

In a connected graph G, a cut set refers to a set S of edges characterized by the following properties:

- **The** removal of all edges in S results in the disconnection of G.

- However, the removal of only some, but not all, edges in S does not disconnect G.

**Example 1**

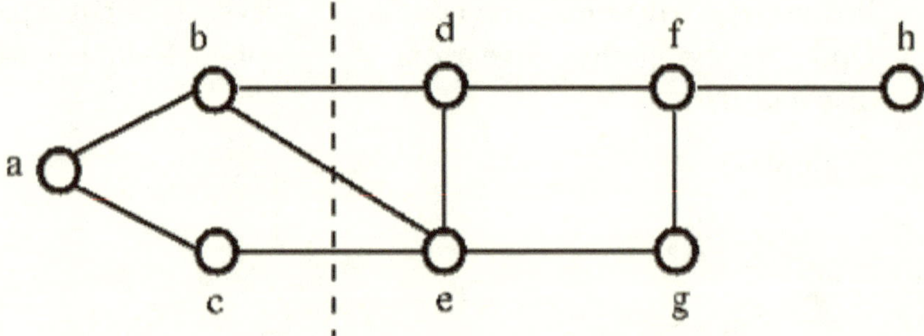

In order to disconnect the depicted graph G, we need to remove three edges: bd, be, and ce. Merely removing two out of these three edges does not suffice to disconnect the graph

Therefore, {bd, be, ce} constitutes a cut set.

After removing this cut set from the graph, it would appear as follows:

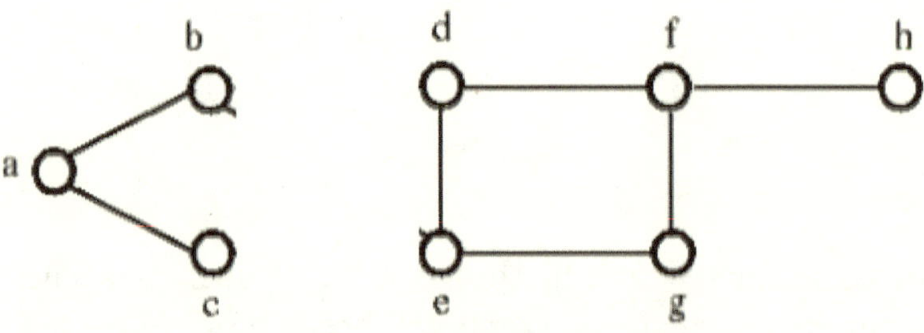

## 4)   Edge Connectivity

The edge connectivity of a connected graph G represents the smallest number of edges that, when removed, disconnect G. This value is denoted by $\lambda(G)$.

When $\lambda(G) \geq k$, then graph G is said to be k-edge-connected.

### Example

Let's see an example,

In the depicted graph, removing the minimum of two edges results in the connected graph becoming disconnected. Therefore, its edge connectivity is 2. Thus, the depicted graph is classified as a 2-edge-connected graph.

Here are four possible methods to disconnect the graph by removing two edges:

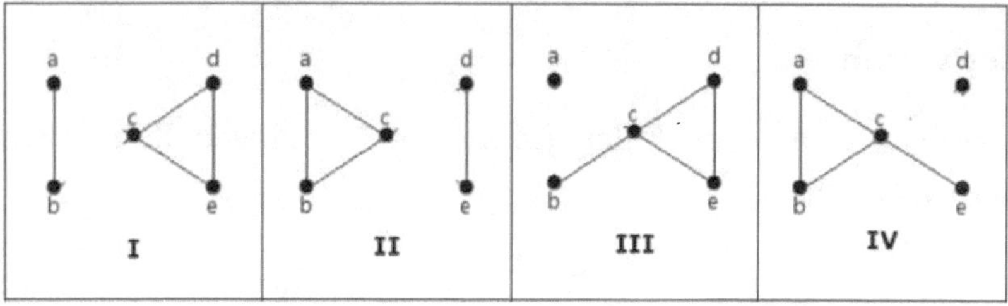

## 5) Vertex Connectivity

The connectivity, also known as vertex connectivity, of a connected graph G denotes the smallest number of vertices whose removal results in the disconnection or reduction of G to a trivial graph. It is symbolized by K(G).

A graph is deemed k-connected or k-vertex connected if its connectivity K(G) is equal to or greater than k. When removing a vertex, we must also eliminate the edges incident to it.

### Example

Let's see an example:

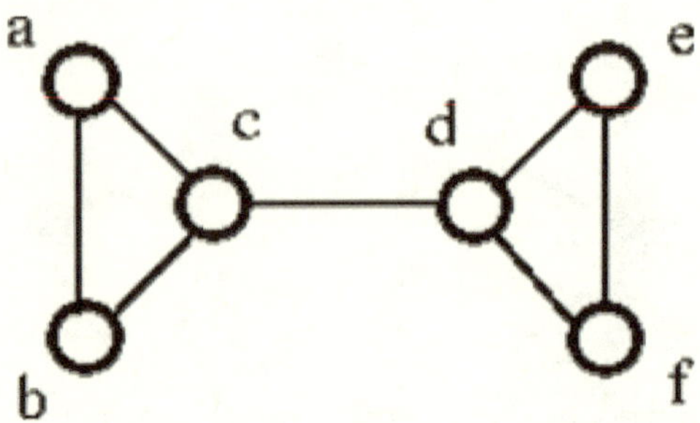

The depicted graph G can be disconnected by removing either the single vertex 'c' or 'd'. Thus, its vertex connectivity is 1, making it a 1-connected graph.

# 4.6. EULER AND HAMILTONIAN PATHS AND CIRCUITS

### Euler Graph

A graph is said to be Euler graph if it has closed Euler Path or Euler Circuit in it.

## Euler Path

It is a path that includes all edges once and vertices may be repeated.

## Closed Euler Path

It is a Euler Path in which the starting and ending vertex are same.

## Note:

a)   If all vertices are of even degree, then it is a Euler graph.

b)   If there are exactly two vertices of odd degree, it is not a Euler Graph, but, there will be a Euler path that will start from one vertex of odd degree and end with other vertex with odd degree.

## Example 1

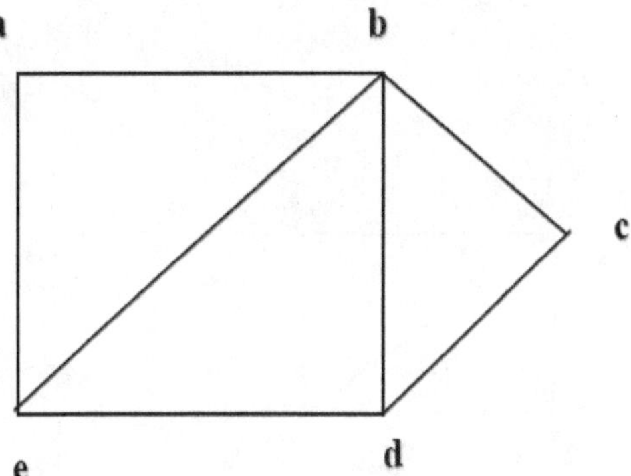


In the above Example 1 graph, the vertices a, b, c are of even degree and the vertices d and e have odd degree. Therefore, this is not a Euler Graph, but there is a Euler path that starts from vertex e and ends in the vertex d. The Euler path is e -> a -> b -> e -> d -> b -> c -> d.

## Example 2

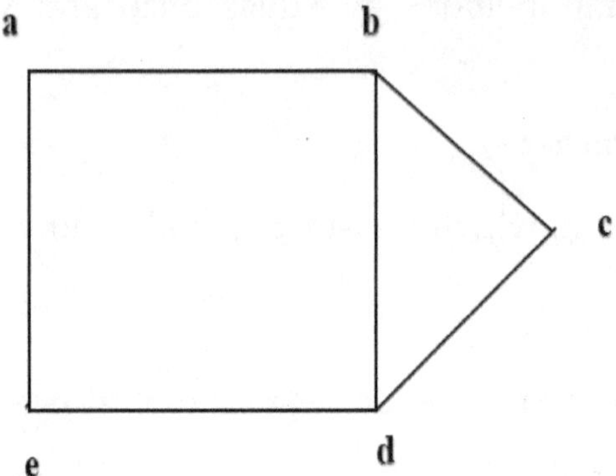



In the above Example 2 graph, the vertices a, e and c have even degree and the vertices b and d have odd degree. Therefore, this is not a Euler Graph, but there is a Euler path that starts from vertex d and ends in vertex b. The Euler path is d -> e -> a -> b -> c -> d -> b.

## Example 3

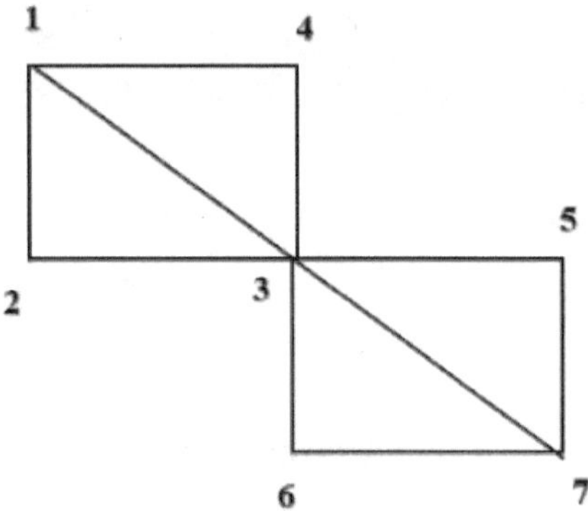



In the above Example 3 graph, the vertices 2, 3, 4, 5 and 6 have even degree and the vertices 1 and 7 have odd degree. Therefore, this is not a Euler Graph, but there is a Euler path that starts from vertex 1 and ends in vertex 7.

The Euler path is 1 -> 4 -> 3 -> 2 -> 1 -> 3 -> 5 -> 7 -> 6 -> 3 -> 7.

**Example 4**

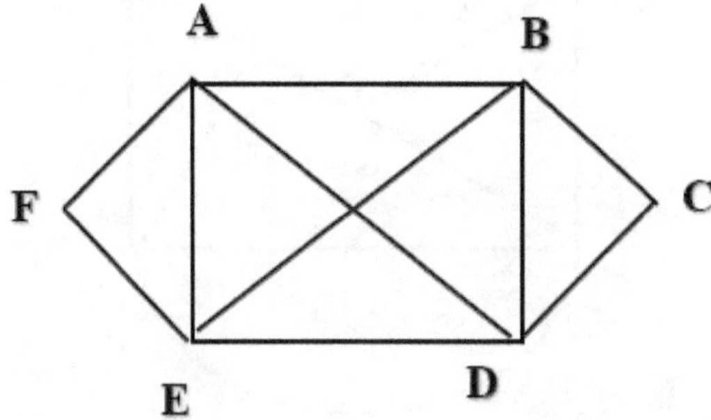

In the above Example 4 graph, all the vertices have even degree and hence it is a Euler graph and it consists of a closed Euler path A -> B -> D -> C -> B -> E -> D -> A -> E -> F -> A

**Example 5**

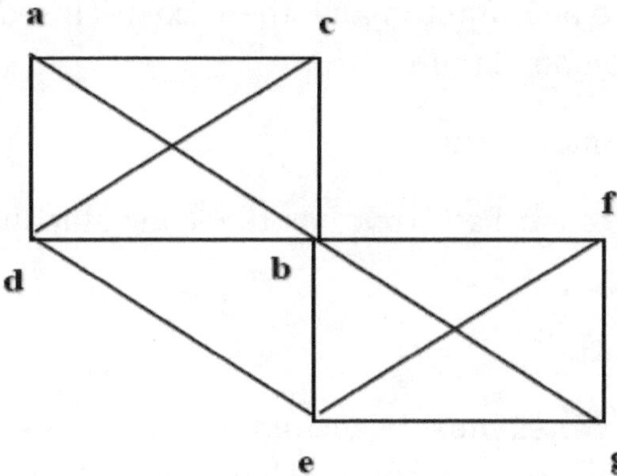

In the above Example 5 graph, not all vertices have even degree and hence it is not a Euler graph. Also, there are 4 vertices (a, c, f, g) with odd degree and hence it will not have a Euler path as well.

## Example 6

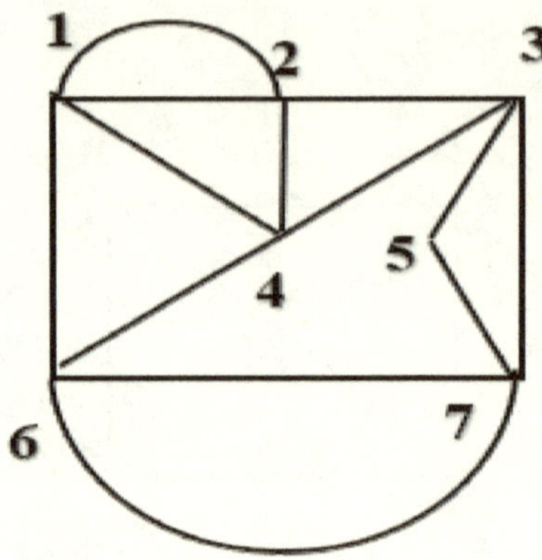

In the above Example 6 graph, all vertices are of even degree and hence it is a Euler graph with the closed Euler path 6 -> 7 -> 6 -> 4 -> 3 -> 5 -> 7 -> 3 -> 2 -> 1 -> 2 -> 4 -> 1 -> 6.

## Hamiltonian Graph

A graph is said to be Hamiltonian if there exists closed Hamiltonian Path or Hamiltonian Circuit.

## Closed Hamiltonian Path

It is the Hamiltonian Path that has the same starting and ending vertex.

## Hamiltonian Path

Each and every vertex must be visited once (except starting vertex) and no restriction on the edges. That is edges can be traversed repeatedly.

**Example 1**

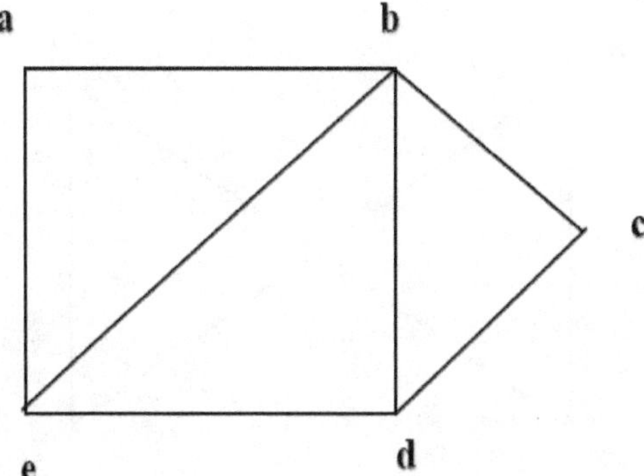

In the above example 1 graph, we can see it is a Hamiltonian graph with the Hamiltonian circuit a -> b -> c -> d -> e -> a.

**Example 2**

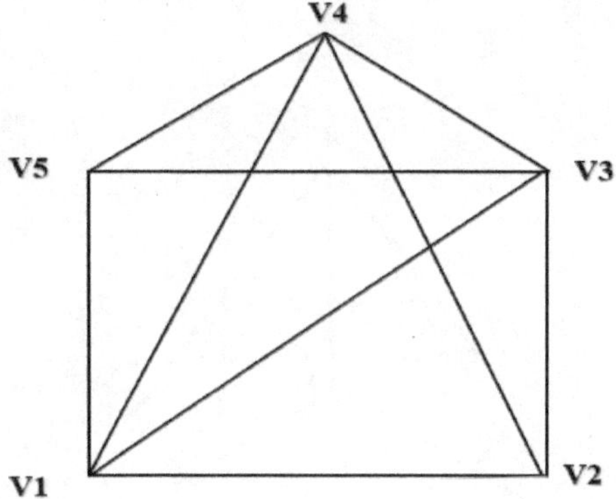

In the above example 2 graph, we can see it is a Hamiltonian graph with the Hamiltonian circuit V1 -> V2 -> V3 -> V4 -> V5 -> V1.

## Example 3

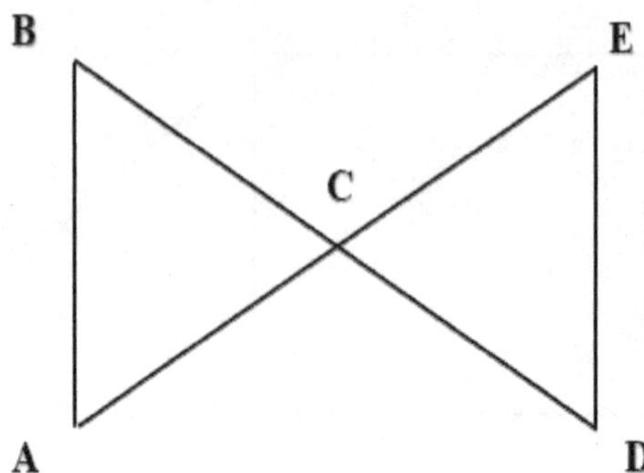


In the above example 3 graph, we can see that there exists a Hamiltonian Path A -> B -> C -> D, but not a Hamiltonian circuit as the vertex A cannot be reached from the vertex D without passing through the vertex C. Hence this is not a Hamiltonian graph, but there exists a Hamiltonian path.

## Example 4

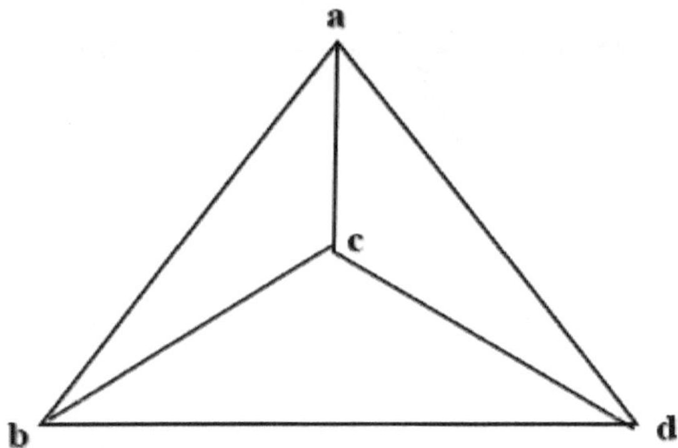


In the above example 4 graph, we can see it is a Hamiltonian graph with a Hamiltonian circuit a -> b -> c -> d -> a.

## Example 5

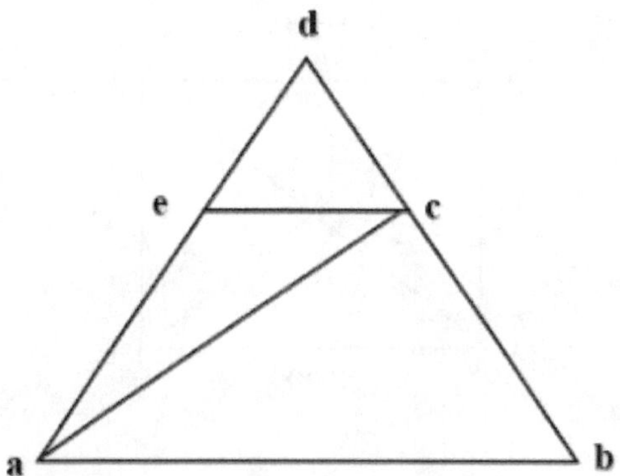

In the above example 5 graph, we can see that a Hamiltonian path a -> b -> c -> e -> d exists, but there is no Hamiltonian circuit. That is the vertex a can not be reached from vertex d without passing through the vertex e. Hence, this is not a Hamiltonian graph, but a Hamiltonian path exists.

## Example 6

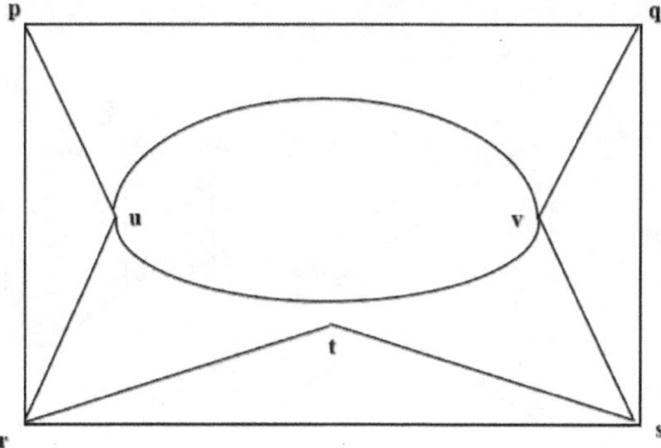

In the above example 6 graph it has a Hamiltonian path p -> u -> v -> q -> s -> r -> t, but it is not a Hamiltonian graph.

## Example of Euler and Hamiltonian Graph

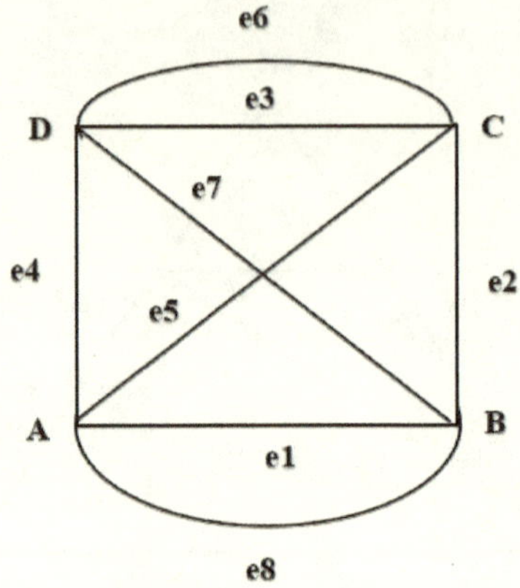


**Closed Euler path:**

A -> e1 -> B -> e2 -> C -> e3 -> D -> e4 -> A -> e5 -> C ->
e6 -> D -> e7 -> B -> e8 -> A

**Closed Hamiltonian path:**

A -> e1 -> B -> e2 -> C -> e3 -> D -> e4 -> A

## Example of Euler but not a Hamiltonian Graph

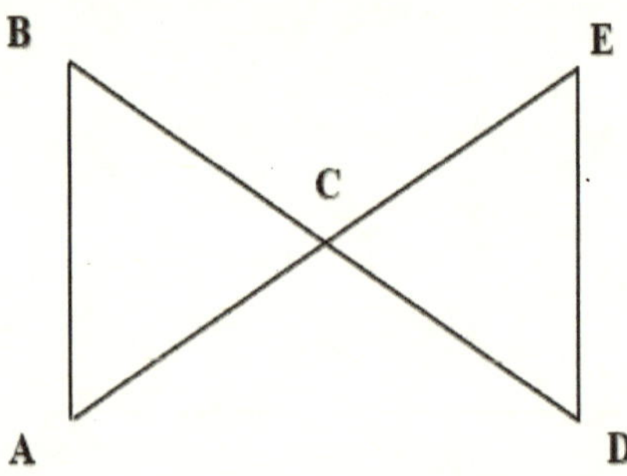


**Closed Euler path:**

A -> B -> C -> E -> D -> C -> A

## Hamiltonian path:

$$A \to B \to C \to D \to E$$

## Example of Hamiltonian but not a Euler Graph

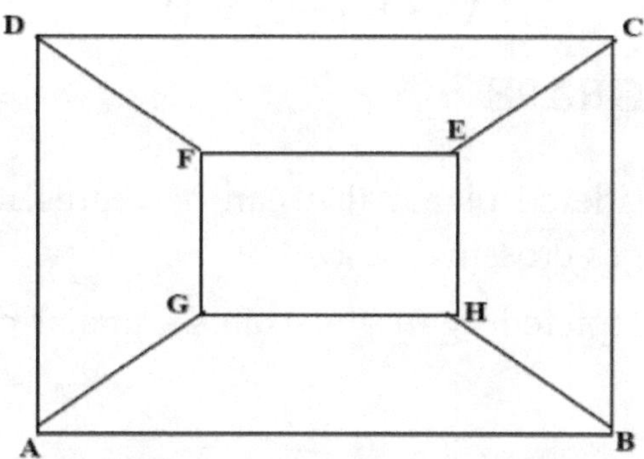

Not a Euler graph as the vertices do not have even degree. No Euler path also exists.

## Closed Hamiltonian path:

$$A \to B \to C \to D \to F \to E \to H \to G \to A$$

## Example of a neither Euler nor a Hamiltonian Graph

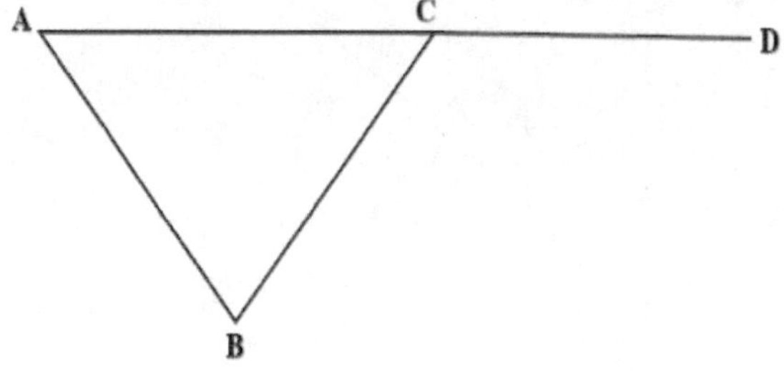

**Euler path:**

$$D \to C \to A \to B \to C$$

**Hamiltonian path:**

$$A \to B \to C \to D$$

## 4.7. PLANAR GRAPH

A graph is considered planar if it can be represented in a plane without any edges crossing.

**Example:** The depicted figure illustrates a planar graph.

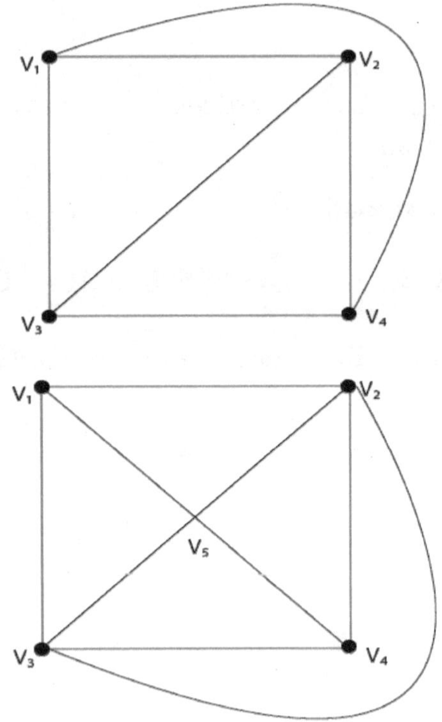

**Region of a Graph**: In the context of a planar graph $G=(V,E)$, a region refers to a distinct area of the plane bounded by edges and incapable of further subdivision. A planar graph partitions the plane into one or more regions, one of which is infinite.

**Finite Region:** A region is termed finite if its area is finite.

**Infinite Region:** A region is deemed infinite if its area is infinite. A planar graph possesses only one infinite region.

**Example:** Examine the graph depicted in the figure to identify the count of regions, finite regions, and an infinite region.

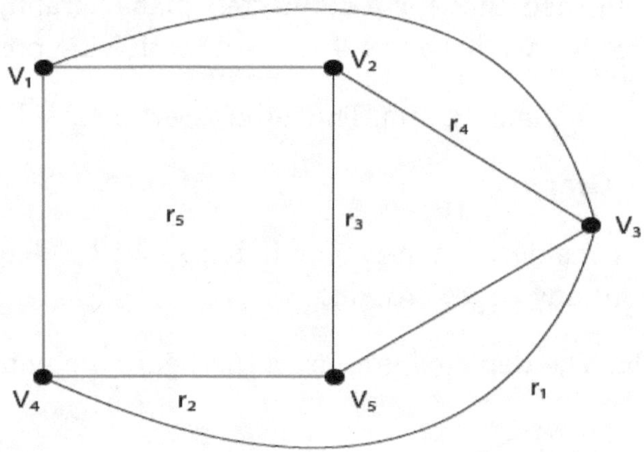


**Solution:** There are five regions in the above graph, i.e. $r_1, r_2, r_3, r_4, r_5$.

There are four finite regions in the graph, i.e., $r_2, r_3, r_4, r_5$.

There is only one finite region, i.e., $r_1$

**Properties of Planar Graphs:**

- For a connected planar graph G with e edges and r regions, the inequality $r \leq (2/3) e$ holds.

- For a connected planar graph G with e edges, v vertices, and r regions, the equation $v - e + r = 2$ is satisfied.

- For a connected planar graph G with e edges and v vertices, the inequality $3v - e \geq 6$ is valid.

- A complete graph $K_n$ is planar if and only if n < 5.

- A complete bipartite graph $K_{mn}$ is planar if and only if m < 3 or n > 3.

**Example:** Demonstrate that the complete graph $K_4$ is planar.

**Solution:** The complete graph $K_4$ consists of 4 vertices and 6 edges.

It's established that for a connected planar graph, $3v - e \geq 6$. Hence, for $K_4$, we have $3 \times 4 - 6 = 6$, satisfying property (3).

Thus, $K_4$ is a planar graph. Thus, the assertion is substantiated.

**Non-Planar Graph:**

A graph is considered non-planar if it cannot be depicted in a plane without any edges crossing.

**Example:** The depicted graphs in the figure are non-planar.

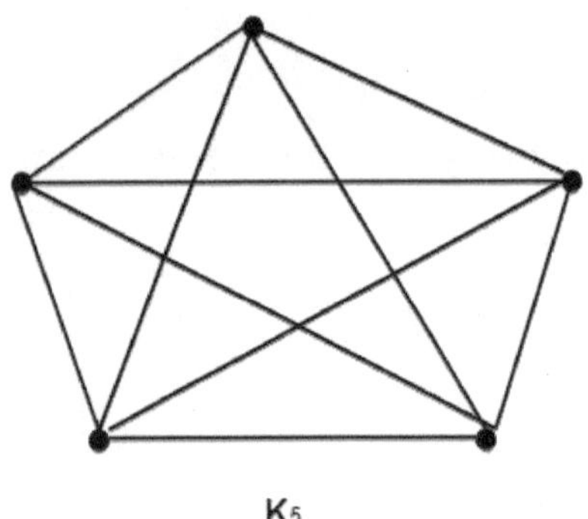


These graphs cannot be represented in a plane without edges crossing, thus classifying them as non-planar graphs.

**Example1:** Demonstrate that $K_5$ is non-planar.

**Solution:** The complete graph $K_5$ comprises 5 vertices and 10 edges.

Now, for a connected planar graph, $3v - e \geq 6$.

Thus, for $K_5$, we have $3 \times 5 - 10 = 5$, which does not fulfill property 3 since it must be greater than or equal to 6.

Consequently, $K_5$ is a non-planar graph.

## 4.8. GRAPH COLORING

Graph coloring involves assigning colors to the vertices of a graph, ensuring that adjacent vertices do not share the same color. This process is also referred to as Vertex Coloring. In graph coloring, it is crucial to ensure that no edge connects vertices of the same color. A graph meeting this criterion is termed a Properly colored graph.

## Example of Graph coloring

This graph depicts a properly colored graph, defined as follows:

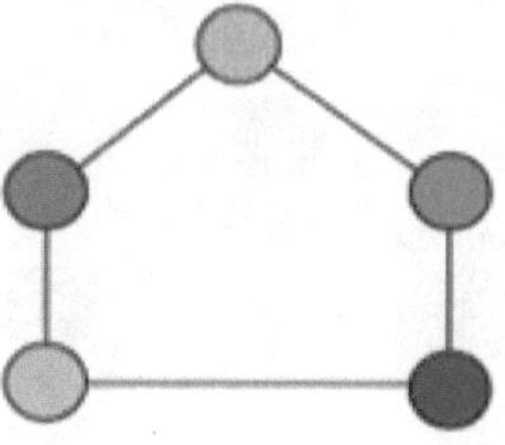

The depicted graph features certain points, elucidated as follows:

- Adjacent vertices cannot be colored with the same color.
- Thus, it can be designated as a properly colored graph.

## Applications of Graph coloring

Graph coloring finds application in various contexts. Here are some of its significant applications:

- Assignment
- Map coloring
- Scheduling the tasks
- Sudoku
- Prepare time table
- Conflict resolution

## Chromatic Number

The chromatic number refers to the smallest number of colors necessary to properly color any graph. Essentially, it represents the minimum number of colors required to color a graph in such a manner that no two adjacent vertices share the same color.

## Example of Chromatic number:

To grasp the concept of the chromatic number, let's examine a graph outlined below:

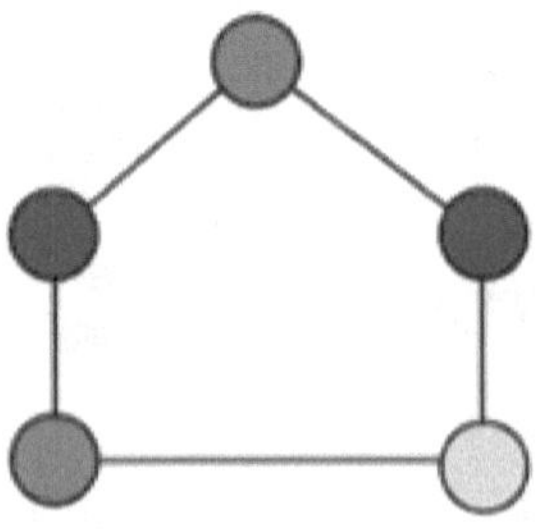

The depicted graph has certain characteristics, outlined as follows:

- Adjacent vertices are assigned distinct colors.
- The minimum number of colors needed to properly color the vertices in this graph is 3.
- Thus, the chromatic number of this graph is 3.
- To properly color this graph, a minimum of 3 colors is required.

## Types of Chromatic Number of Graphs:

Various types of chromatic numbers for graphs are delineated below:

## Cycle Graph:

A graph is classified as a cycle graph if it comprises 'n' edges and 'n' vertices (n ≥ 3), forming a cycle of length 'n'. The degrees of all vertices in the cycle graph can only be 2 or 3.

## Chromatic number:

- The chromatic number in a cycle graph is 2 when the number of vertices in the graph is even.
- The chromatic number in a cycle graph is 3 when the number of vertices in the graph is odd.

## Examples of Cycle graphs:

Several examples of cycle graphs exist. Here are some illustrations:

**Example 1:**

In the subsequent graph, we aim to ascertain the chromatic number.

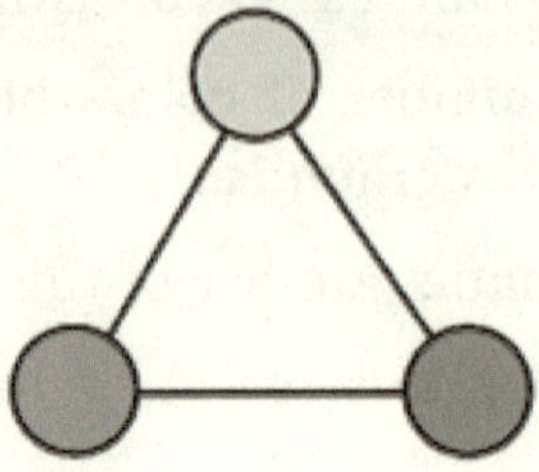

**Solution:**

In the depicted cycle graph, three different colors are assigned to three vertices, and no adjacent vertices share the same color. Since the number of vertices in this graph is odd, the Chromatic number is 3.

**Example 2:**

In the subsequent graph, the objective is to ascertain the chromatic number.

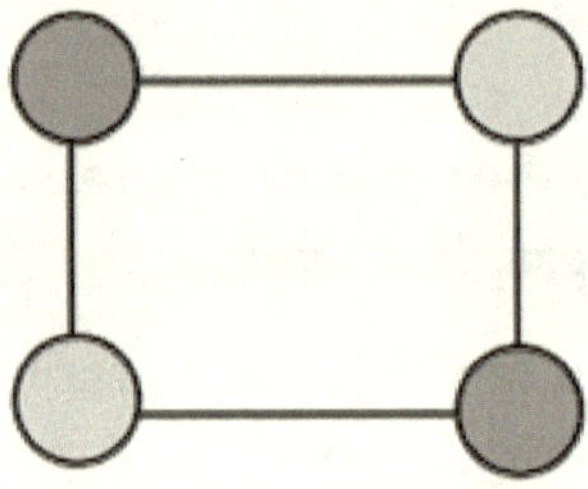

**Solution:**

In the depicted cycle graph, two colors are used for four vertices, and no adjacent vertices share the same color. Since the number of vertices in this graph is even, the Chromatic number is 2.

**Example 3:**

In the subsequent graph, the task is to determine the chromatic number.

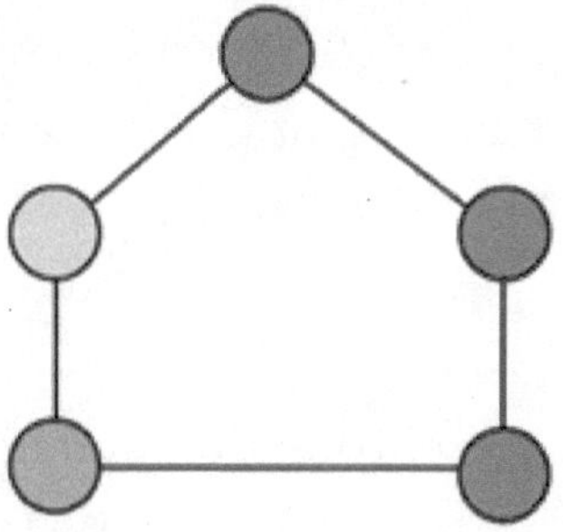

**Solution:**

In the depicted graph, five vertices are colored with four different colors, and two adjacent vertices are assigned the same color (blue). Hence, this graph is not a cycle graph and does not possess a chromatic number.

**Example 4:**

In the subsequent graph, the task is to determine the chromatic number.

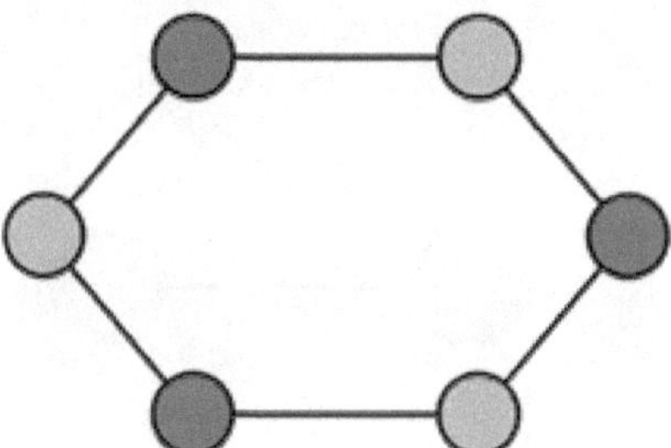

**Solution:**

In the depicted graph, six vertices are colored with two different colors, and no adjacent vertices share the same color. Since the number of vertices in this graph is even, the Chromatic number is 2.

**Planar Graph:**

A graph is termed a planar graph if it can be represented in a plane without any edges crossing.

**Chromatic Number:**

- In a planar graph, the chromatic number must be less than or equal to 4.

- A planar graph can be illustrated by all the aforementioned cycle graphs except example 3.

**Examples of Planar Graphs:**

Several examples of planar graphs exist. Here are some illustrations:

**Example 1:**

In the subsequent graph, the task is to determine the chromatic number.

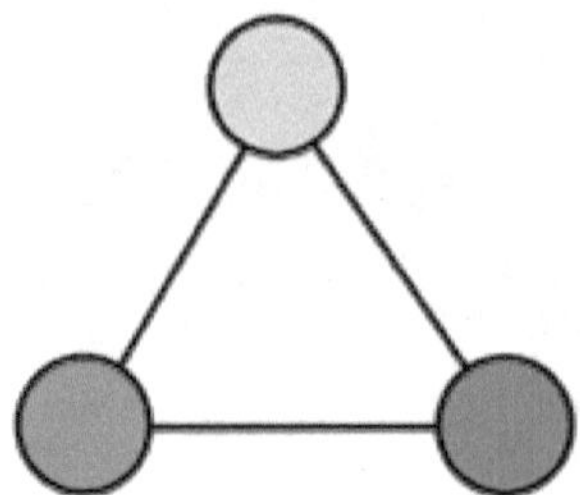

**Solution:**

In the depicted graph, three different colors are assigned to three vertices, and no edges cross each other. Thus, the chromatic number is 3. Since the chromatic number is less than 4, this graph qualifies as a planar graph.

**Example 2:**

In the subsequent graph, the objective is to determine the chromatic number.

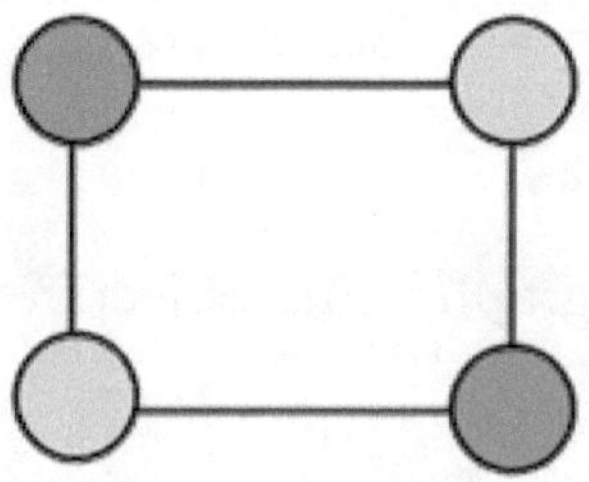

**Solution:**

In the depicted graph, four vertices are colored with two different colors, and no edges cross each other. Hence, the chromatic number is 2. Since the chromatic number is less than 4, this graph qualifies as a planar graph.

**Example 3:**

In the subsequent graph, the task is to determine the chromatic number.

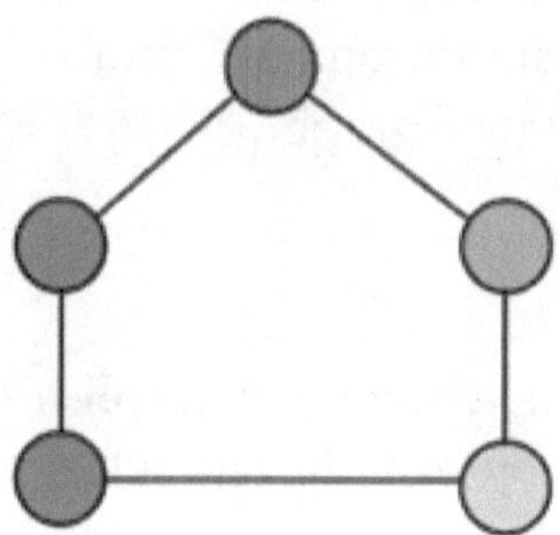

**Solution:**

In the depicted graph, five vertices are colored with five different colors, and no edges cross each other. Hence, the chromatic number is 5. Since the chromatic number is greater than 4, this graph does not qualify as a planar graph.

**Example 4:**

In the subsequent graph, the objective is to determine the chromatic number.

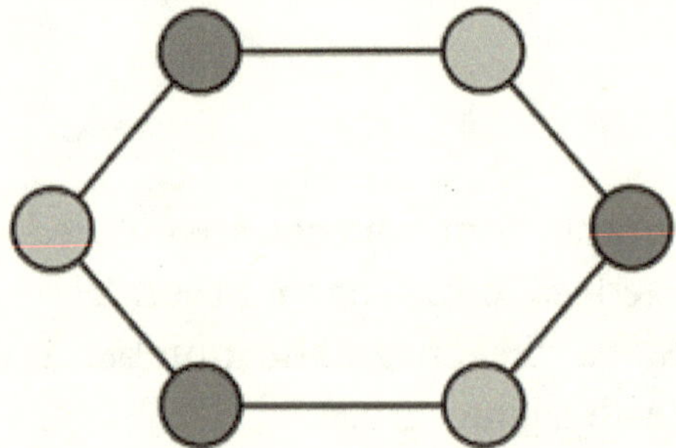

**Solution:**

In the depicted graph, six vertices are colored with two different colors, and no edges cross each other. Hence, the chromatic number is 2. Since the chromatic number is less than 4, this graph qualifies as a planar graph.

## Complete Graph:

A graph is classified as a complete graph if every pair of distinct vertices is connected by exactly one edge. In a complete graph, each vertex is linked to every other vertex. Consequently, every vertex in a complete graph is assigned a distinct color, ensuring that no two vertices share the same color.

## Chromatic Number:

In a complete graph, the chromatic number is equivalent to the number of vertices in that graph.

## Examples of Complete Graphs:

Several examples of complete graphs exist. Here are some illustrations:

**Example 1:**

In the subsequent graph, the task is to determine the chromatic number.

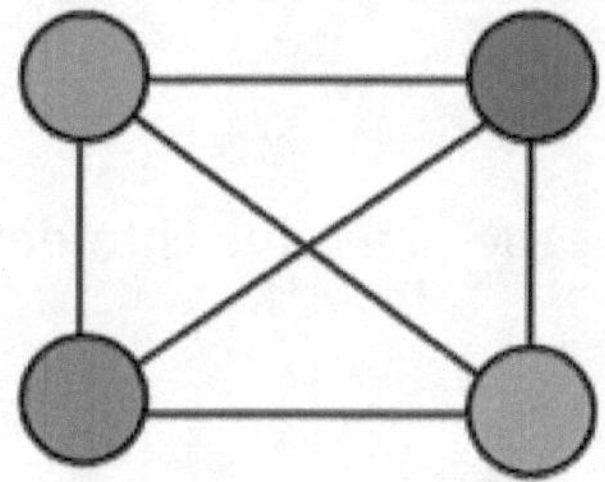

**Solution:**

In the depicted graph, four different colors are assigned to four distinct vertices, and no two colors are identical. According to the definition, the chromatic number equals the number of vertices. Therefore, the Chromatic number is 4.

**Example 2:**

In the subsequent graph, the objective is to determine the chromatic number.

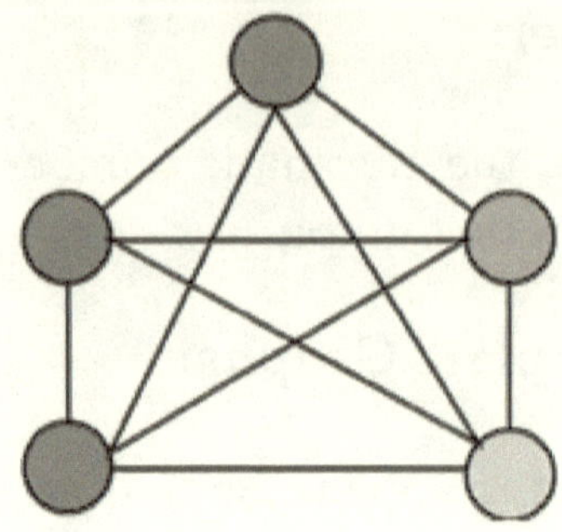

**Solution:**

In the provided graph, five unique vertices are assigned five distinct colors, ensuring no repetition of colors. According to the definition, the chromatic number is equivalent to the number of vertices. Thus, the chromatic number is 5.

**Example 3:**

In the following graph, the objective is to determine the chromatic number.

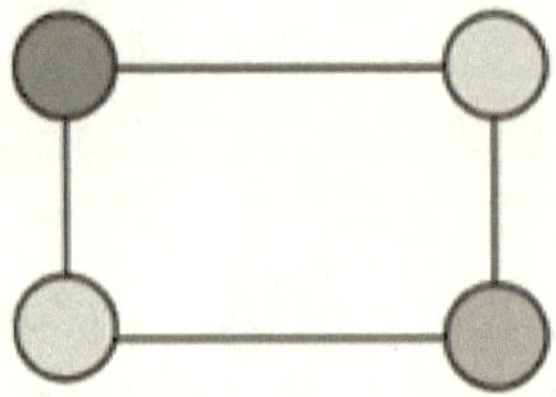

**Solution:**

In the provided graph, three different colors are assigned to four distinct vertices, with one color repeated in two vertices. Consequently, this graph does not qualify as a complete graph and does not have a chromatic number.

## Bipartite Graph:

A graph is classified as a bipartite graph if its vertices can be divided into two disjoint sets, denoted as A and B, such that every edge connects a vertex from set A to a vertex from set B.

## Chromatic Number:

In any bipartite graph, the chromatic number is always 2.

## Examples of Bipartite Graphs:

Numerous examples of bipartite graphs are available. Here is an example:

**Example 1:**

In the subsequent graph, the objective is to determine the chromatic number.

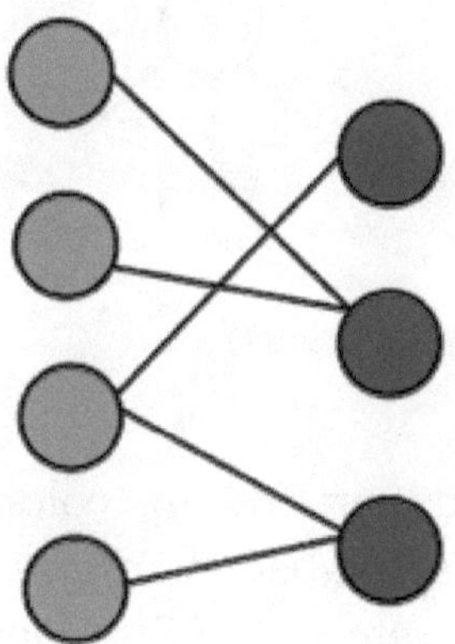

**Solution:**

In the depicted graph, two distinct sets of vertices are evident. Consequently, the chromatic number of all bipartite graphs will consistently remain 2. Thus, the Chromatic number is 2.

## Tree:

A graph is classified as a tree if it is connected and contains no cycles. In a tree, irrespective of the number of vertices it possesses, the chromatic number always equals 2. Furthermore, every bipartite graph is also considered a tree.

## Chromatic Number:

In any tree, the chromatic number is always 2.

## Examples of Trees:

Numerous examples of trees exist. Here is one:

**Example 1:**

In the subsequent tree, the task is to determine the chromatic number.

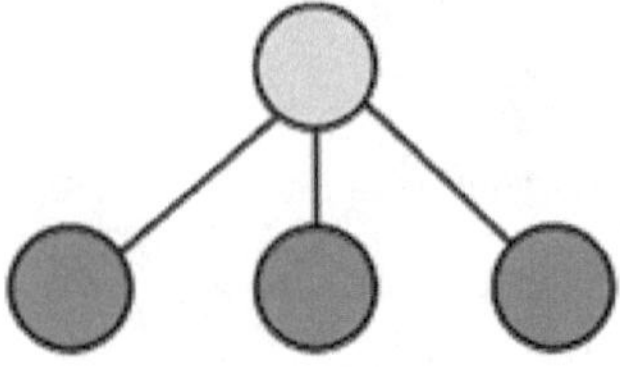

**Solution:**

In the given tree, four vertices are colored using two distinct colors. Regardless of the number of vertices in a tree, the chromatic number remains 2. Therefore, the Chromatic number is 2.

**Example 2:**

In the subsequent tree, the objective is to determine the chromatic number.

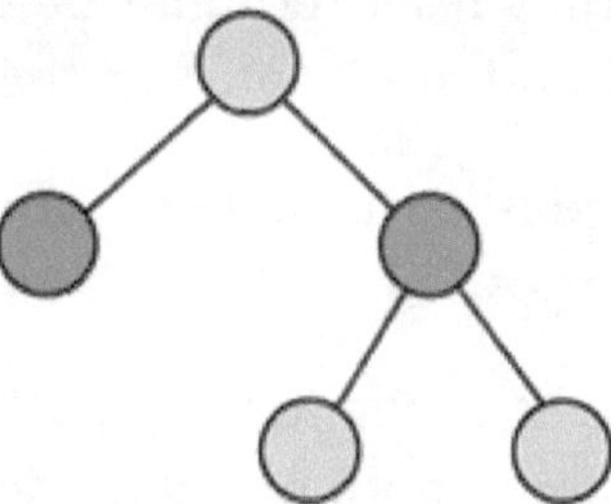

**Solution:**

There are 2 different colors for five vertices. A tree with any number of vertices must contain the chromatic number as 2 in the above tree. So, Chromatic number = 2

## 4.9. TREES BASIC TERMINOLOGY

In discrete structures, trees serve as fundamental data structures that represent hierarchical relationships or connections between elements. They find extensive applications in computer science, mathematics, and diverse fields due to their simplicity and versatility.

**Definition:** In graph theory, a tree is an undirected graph where any two vertices are connected by exactly one path. Put differently, a tree is a connected graph without any cycles.

**Nodes and Edges:** Trees represent elements as nodes, with connections between nodes represented by edges. Typically, each node has a parent node (except the root) and zero or more child nodes.

**Root:** The root is the topmost node in a tree, devoid of any parent nodes. All other nodes descend from the root.

**Parent and Child Nodes:** Nodes with one or more child nodes are termed parent nodes, while child nodes are those with parent nodes.

**Leaf Nodes:** Also known as terminal or external nodes, leaf nodes have no children and are located at the tree's bottom.

**Internal Nodes:** Nodes with at least one child are internal nodes, excluding leaf nodes.

**Height:** A tree's height is the longest path from the root to a leaf node, or the maximum depth of any node.

**Depth:** The depth of a node is the length of the path from the root to that node, with the root having a depth of 0.

**Binary Trees:** Binary trees limit each node to at most two children, typically referred to as left and right children. They are extensively used in data storage and searching algorithms.

**Traversal:** Traversal involves visiting all nodes in a tree in a specific order, commonly done through depth-first (pre-order, in-order, post-order) or breadth-first (level-order) traversal.

**Balanced Trees:** These maintain efficient operations like search, insertion, and deletion by ensuring that subtree heights differ by at most one.

**Ancestor and Descendant:** An ancestor of a node is any node on the path from the root to that node, while a descendant is reachable by traversing downward from a node.

**Sibling:** Siblings share the same parent node.

**Subtree:** A subtree comprises a node and all its descendants, forming a smaller tree within the larger one.

**Forest:** A collection of disjoint trees forms a forest, with no common nodes or edges.

**Path:** A path in a tree is a sequence of nodes connected by edges, starting and ending at specific nodes.

**Rooted Tree:** This designates one node as the root, distinguishing it from general trees.

**Complete Binary Tree:** Every level, except possibly the last, is fully filled, and the last level fills from left to right.

**Perfect Binary Tree:** Internal nodes have two children, and leaf nodes are at the same level.

**Traversal Path:** The order in which nodes are visited during traversal algorithms.

**Applications:** Trees are used for hierarchical data representation, implementing data structures and algorithms, among other applications.

## 4.10. PROPERTIES OF TREES

**Connectedness:** A tree is a connected graph, meaning that there is a path between every pair of nodes in the tree. This property ensures that all elements in the tree are accessible from one another.

**Acyclicity:** Trees are acyclic graphs, which means they contain no cycles. In other words, there are no sequences of edges that form closed loops within the tree. This property ensures that there is only one unique path between any two nodes in the tree.

**Uniqueness of Paths:** In a tree, there is exactly one unique path between any pair of nodes. This property is a consequence of the acyclicity property and ensures that trees have well-defined hierarchical structures.

**Node Relationships:** Each node in a tree (except the root) has exactly one parent node, and zero or more child nodes. This hierarchical parent-child relationship is a fundamental characteristic of trees.

**Number of Nodes and Edges**: If a tree has n nodes, it has $n-1$ edges. This property is known as the "one less than the number of nodes" property and holds true for all trees.

**Degree:** The degree of a node in a tree is the number of children it has. In a binary tree, nodes have at most two children, resulting in a maximum degree of 2.

**Unique Path:** As mentioned before, trees have a unique path between any pair of nodes. This property implies that there are no alternate paths between two nodes, which simplifies navigation within the tree structure.

**No Parallel Edges:** In a tree, there are no parallel edges, meaning that each pair of nodes is connected by exactly one edge. This property distinguishes trees from general graphs, where multiple edges between the same pair of nodes are allowed.

**Finite Structure:** Trees have a finite number of nodes and edges. This finite structure makes trees suitable for representing hierarchical relationships in various applications, including computer science, biology, and organizational structures.

**Recursive Structure:** Trees exhibit a recursive structure, where each subtree is itself a tree. This property allows for the application of recursive algorithms to traverse, search, and manipulate tree structures efficiently.

**Ordered Structure:** In some types of trees, such as binary trees and binary search trees, the arrangement of child nodes relative to their parent node is significant. For instance, in a binary tree, each node has a left child and a right child, and the order of insertion determines their placement.

**Height Balance (for Balanced Trees):** Balanced trees, such as AVL trees and Red-Black trees, maintain a balance condition to ensure efficient operations. The balance condition typically involves ensuring that the heights of the subtrees of any node differ by at most one, which helps in achieving optimal time complexity for operations like insertion, deletion, and search.

**Parent-Child Relationships:** Trees exhibit clear parent-child relationships between nodes. Each node (except the root) has a parent node, and all nodes descended from the same parent are considered its children. This hierarchical relationship is central to the definition and structure of trees.

**Subtree Property:** Every subtree of a tree is also a tree. This property allows for the decomposition of a tree into smaller subtrees, each of which retains the fundamental properties of a tree. It is useful for analyzing and manipulating tree structures recursively.

**Depth-First Search (DFS) Property:** Trees can be traversed using depth-first search algorithms, which explore each branch of the tree fully before backtracking. DFS is a fundamental technique for tree traversal and is commonly used in various tree-related algorithms and applications.

**Breadth-First Search (BFS) Property:** Similarly, trees can be traversed using breadth-first search algorithms, which explore all nodes at the current depth level before moving to the next depth level. BFS is particularly useful for finding the shortest path between nodes in an unweighted tree.

Understanding these properties is crucial for analyzing the structure and behavior of trees in discrete structures, as well as for designing algorithms and data structures that utilize trees effectively.

## 4.11. INTRODUCTION TO SPANNING TREES

In discrete structures, particularly in graph theory, a spanning tree of a connected graph is a subgraph that includes all the vertices of the original graph while forming a tree (i.e., acyclic and connected). Spanning trees are significant in various contexts, including network design, optimization problems, and algorithmic applications. Here's a closer look at the concept of spanning trees:

**Definition:** A spanning tree T of a connected graph G is a subgraph that includes all the vertices of G and is also a tree. Formally, T is a tree that spans all the vertices of G.

**Connectedness:** A spanning tree must maintain the connectedness of the original graph. This means that there should be a path between any pair of vertices in the spanning tree.

**Acyclicity:** A spanning tree must be acyclic, meaning it cannot contain any cycles. Since trees are defined as connected acyclic graphs, this property ensures that the spanning tree remains a tree.

**Spanning Property:** Every vertex in the original graph must be included in the spanning tree. In other words, the spanning tree "spans" or covers all the vertices of the original graph.

**Minimum Number of Edges:** A spanning tree contains the minimum number of edges required to maintain connectivity among all vertices. In an n-vertex connected graph, a spanning tree will have exactly $n-1$ edges.

**Applications:** Spanning trees have practical applications in network design, where they represent the minimal infrastructure needed to connect all locations in a network. They are also used in various optimization problems, such as finding the minimum-cost spanning tree in a weighted graph.

**Algorithms:** Several algorithms exist for finding spanning trees in graphs, including Prim's algorithm and Kruskal's algorithm. These algorithms efficiently compute a minimum-cost spanning tree for weighted graphs or any spanning tree for unweighted graphs.

**Unique Properties:** While a graph may have multiple spanning trees, certain properties are unique to specific spanning trees. For example, in a weighted graph, there may be multiple spanning trees with different total weights, but there exists a unique minimum-cost spanning tree.

**Tree Properties:** Despite being a subgraph of the original graph, a spanning tree exhibits all the properties of a tree, such as having a unique path between any pair of vertices, no cycles, and $n-1$ edges for an n-vertex tree.

**Minimum Spanning Tree (MST):** In weighted graphs, the minimum spanning tree is a spanning tree with the minimum possible total edge weight among all spanning trees of the graph. Finding the MST is a common problem with numerous real-world applications.

Spanning trees play a crucial role in graph theory and its applications, providing a simplified representation of connectivity while preserving essential structural properties. They serve as the backbone for various algorithms and solutions in network design, optimization, and routing problems.

**What is a spanning tree?**

A spanning tree is a subgraph of an undirected connected graph that encompasses all the vertices with the minimum possible number of edges. Omitting any vertex renders it non-spanning. It lacks cycles and cannot be disconnected.

Comprising $(n-1)$ edges, where 'n' denotes the number of vertices, a spanning tree's edges may or may not carry assigned weights. While the vertices remain constant across all possible spanning trees derived from a given graph G, the number of edges in the spanning tree equals the number of vertices in G minus one.

A complete undirected graph can have $n^{n-2}$ number of spanning trees where n is the number of vertices in the graph. Suppose, if n = 5, the number of maximum possible spanning trees would be $5^{5-2} = 125$.

## Applications of the spanning tree

In essence, a spanning tree serves to establish a minimal path connecting all nodes within a graph. Its utility extends to various applications, including:

- Cluster Analysis
- Civil network planning
- Computer network routing protocols

To illustrate the concept of a spanning tree, let's delve into an example.

## Example of Spanning tree

Suppose the graph be -

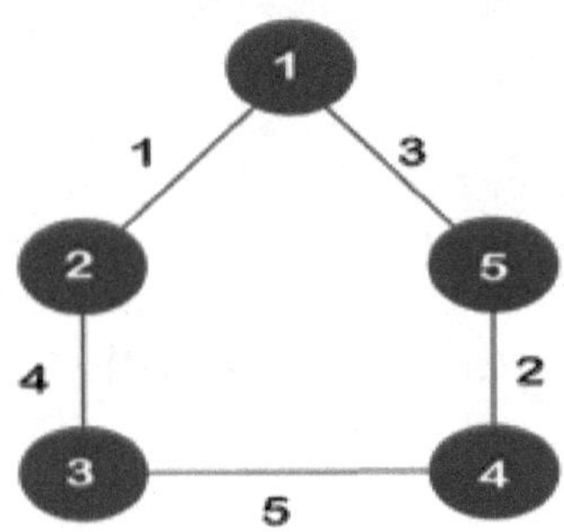


As previously mentioned, a spanning tree encompasses the same count of vertices as the original graph. In the case of the provided graph, which comprises 5 vertices, the spanning tree will also consist of 5 vertices. The number of edges within the spanning tree equals the vertex count of the graph minus 1. Consequently, the spanning tree will contain 4 edges.

Here are several potential spanning trees derived from the given graph:

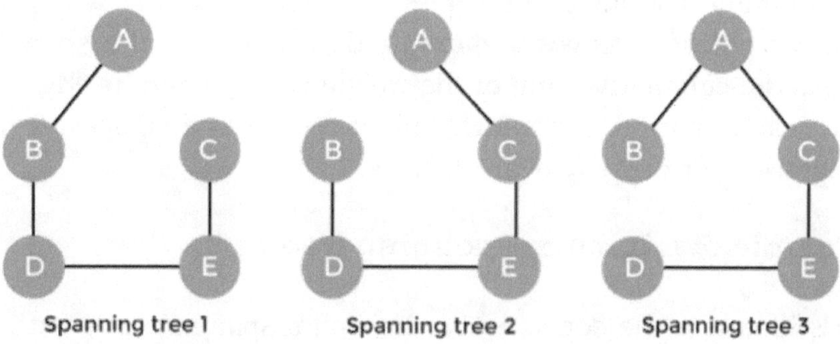



## Properties of spanning-tree

Here are some properties of a spanning tree:

- There can be multiple spanning trees for a connected graph G.

- A spanning tree is cycle-free and doesn't contain any loops.

- It is minimally connected, meaning that removing any single edge from it would disconnect the graph.

- A spanning tree is maximally acyclic; adding any single edge to it would introduce a cycle.

- The maximum number of spanning trees that can be formed from a complete graph is $n^{n-2}$.

- A spanning tree has $n-1$ edges, where 'n' represents the number of nodes.

- If the graph is complete, a spanning tree can be obtained by removing a maximum of $(e-n+1)$ edges, where 'e' denotes the number of edges and 'n' is the number of vertices.

Therefore, a spanning tree is a subset of a connected graph G, and there is no spanning tree for a disconnected graph.

## Minimum Spanning tree

A minimum spanning tree refers to a spanning tree where the total weight of its edges is minimized. The weight of a spanning tree is the cumulative sum of the weights assigned to its edges. In practical scenarios, this weight might represent distances, traffic congestion, or other relevant factors.

## Example of minimum spanning tree

Let's illustrate the concept of a minimum spanning tree with an example.

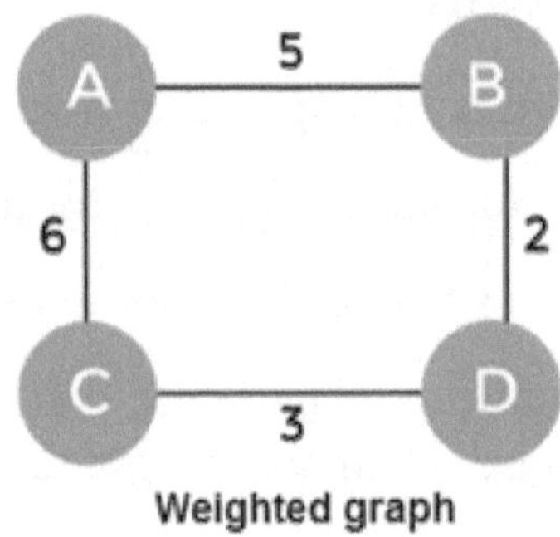

Weighted graph

The total sum of the edges in the depicted graph is 16. Now, we can explore several possible spanning trees derived from this graph.

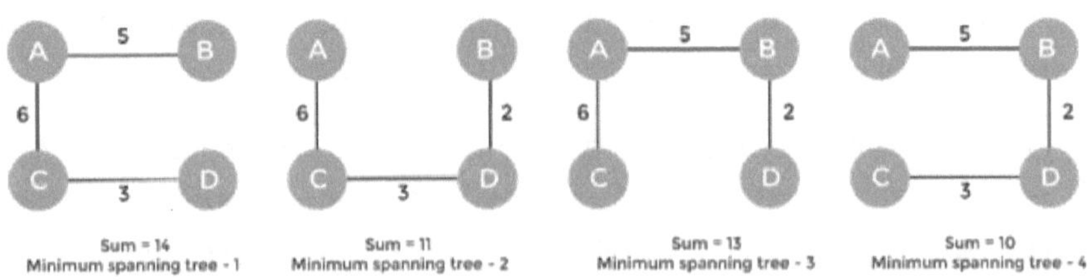


Therefore, the chosen minimum spanning tree from the aforementioned options for the provided weighted graph is:

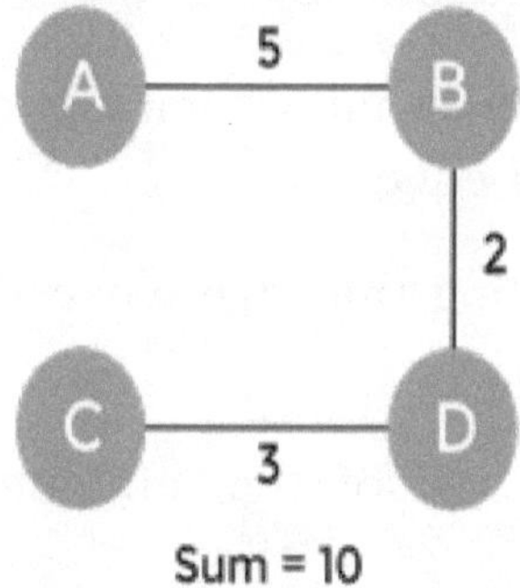


## Applications of minimum spanning tree

The uses of the minimum spanning tree are outlined below:

- Designing water-supply networks, telecommunication networks, and electrical grids.

- Mapping out paths on maps.

## Algorithms for Minimum spanning tree

A minimum spanning tree can be found from a weighted graph by using the algorithms given below -

- Prim's Algorithm

- Kruskal's Algorithm

Let's see a brief description of both of the algorithms listed above.

**Prim's algorithm** - It is a greedy algorithm that starts with an empty spanning tree. It is used to find the minimum spanning tree from the graph. This algorithm finds the subset of edges that includes every vertex of the graph such that the sum of the weights of the edges can be minimized.

Step 1: Select the edge with minimum cost in the entire graph

Step 2: Select the next minimum cost edge that is connected to the already selected edge

Step 3: Stop selecting the edge when the number of edges is number of vertices - 1.

Step 4: Find the cost of the minimum spanning tree thus obtained

Lets see an example of drawing a minimum spanning tree using Prim's Algorithm

**Consider the below graph:**

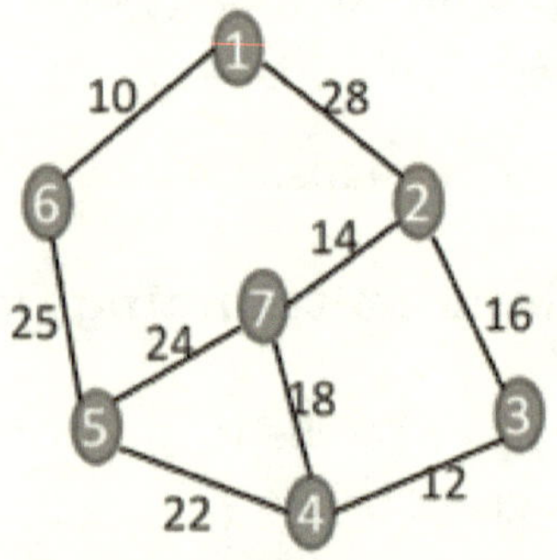


Following Step 1 above we get the below - the edge with minimum cost in the entire graph.

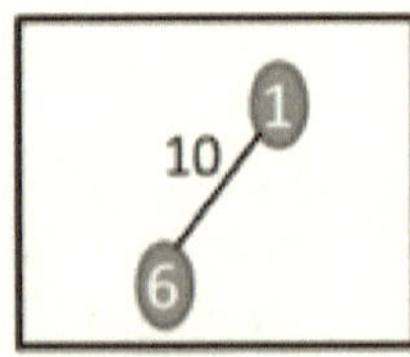


Following Step 2 we get the below graph - the next minimum cost edge that is connected to the already selected edge.

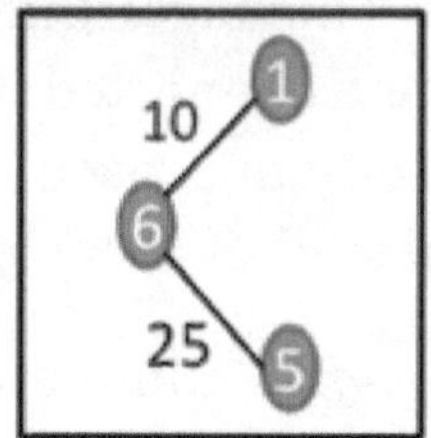

Continue following the Step 2 again we get the below graphs one after the other:

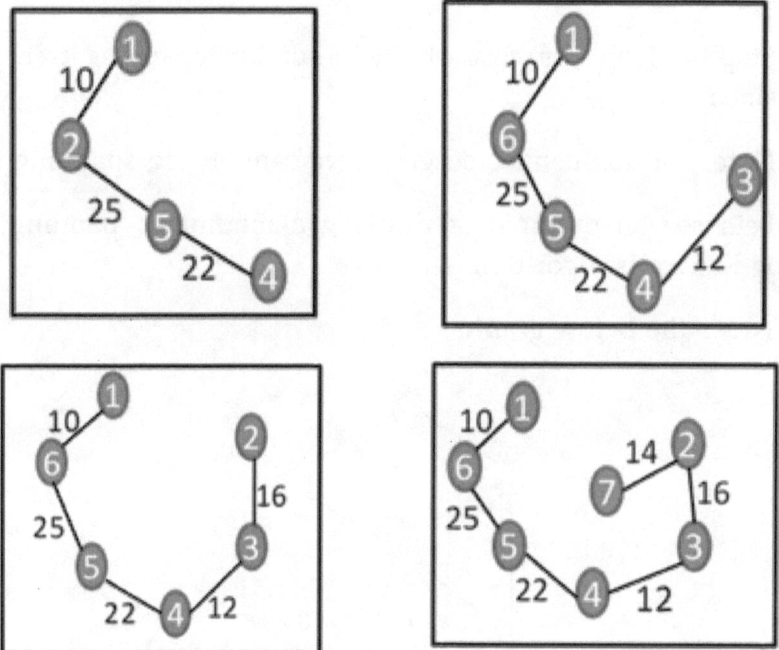

Here we need to stop as per Step 3 - Stop selecting the edge when the number of edges is number of vertices - 1 (7 Vertices -1 = 6 edges). Hence we got the minimum spanning tree as per Prim's Algorithm.

The cost of this minimum spanning tree is 10 + 25 + 22 + 12 + 16 + 14 = 99

**Kruskal's algorithm** - This algorithm is also used to find the minimum spanning tree for a connected weighted graph. Kruskal›s algorithm also follows greedy approach, which finds an optimum solution at every stage instead of focusing on a global optimum.

Step 1: Select the edge with minimum cost in the entire graph

Step 2: Select the next minimum cost edge only if it is not forming the cycle

Step 3: Stop selecting the edge when the number of edges is number of vertices - 1.

Step 4: Find the cost of the minimum spanning tree thus obtained

Note: For non-connected graph we cannot find spanning tree

Lets see an example of drawing a minimum spanning tree using Kruskal's Algorithm

**Consider the below graph:**

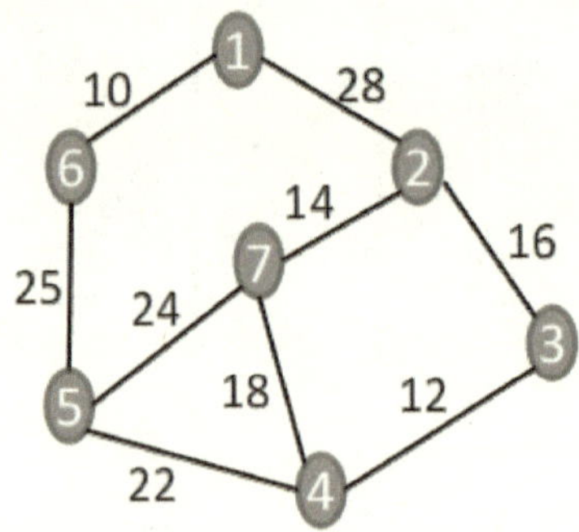


Following Step 1 above we get the below - the edge with minimum cost in the entire graph.

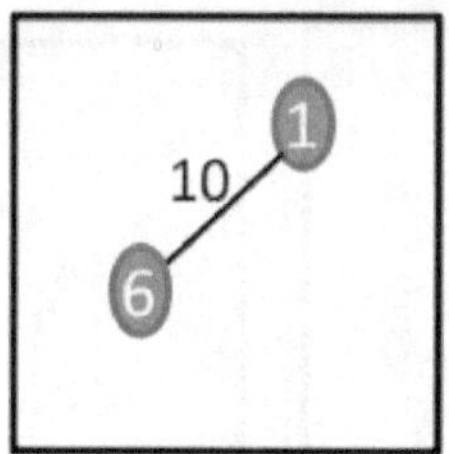

Following Step 2 we get the below graph - the next minimum cost edge only if it is not forming the cycle.

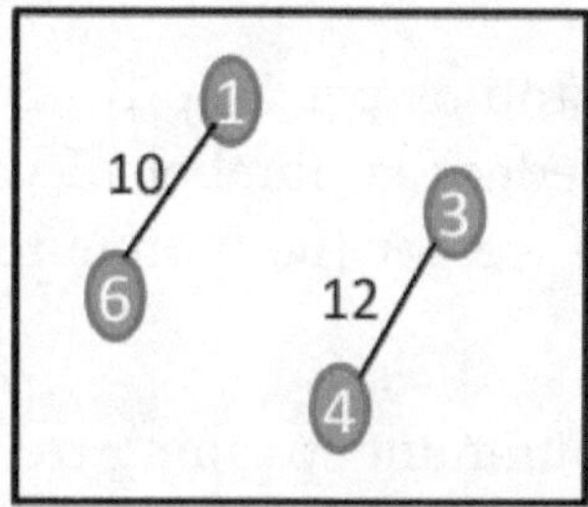

Continue following the Step 2 again we get the below graphs one after the other:

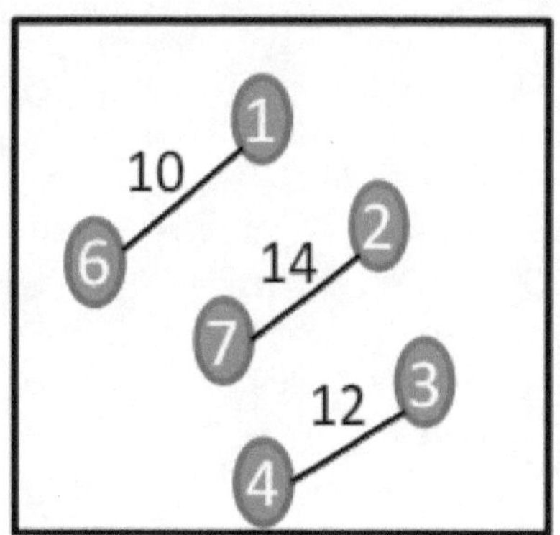

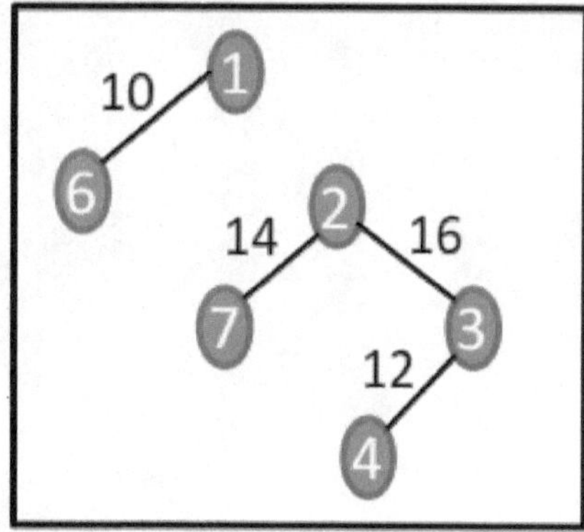

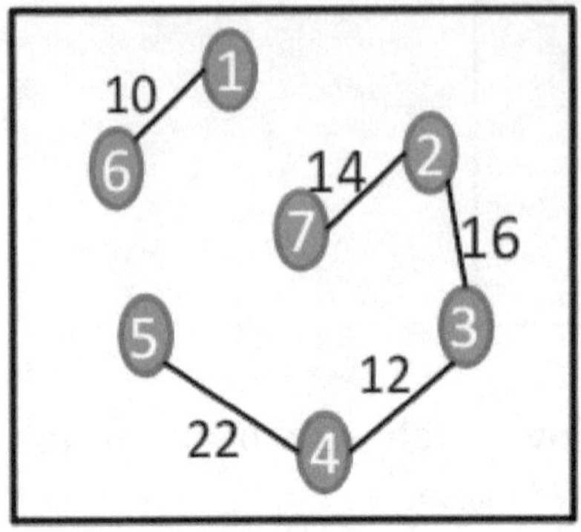 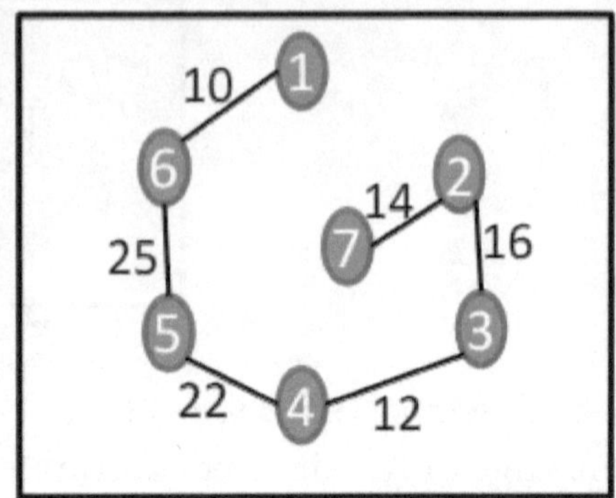

Here we need to stop as per Step 3 - Stop selecting the edge when the number of edges is number of vertices - 1 (7 Vertices -1 = 6 edges). Hence we got the minimum spanning tree as per Kruskal's Algorithm.

The cost of this minimum spanning tree is 10 + 25 + 22 + 12 + 16 + 14 = 99

# Chapter 5

# PROPOSITIONAL LOGIC

## 5.1. PROPOSITIONS

Propositional logic is a basic form of logic employing propositions to construct statements. A proposition, being a declarative statement, asserts facts. Statements in propositional logic are either true or false, but not both simultaneously.

### Examples of Propositions

There exist several instances of propositional logic, exemplified as follows:

- $5 + 2 = 7$
- Bananas are green.
- Yogi Adityanath is the chief minister of Uttar Pradesh.
- Five and five make eight.
- 2021 is the worst year.
- Delhi is the capital of the US.
- Karnataka is in India.

Each of these statements is either true or false, but not both, rendering them propositions.

### Types of Propositions

Propositional logic encompasses two types of propositions, delineated as follows:

1. Atomic Propositions
2. Compound Propositions

## Atomic Propositions

Propositions are termed atomic if they cannot be further divided. These are also referred to as simple propositions, each represented by a single proposition symbol such as p, q, r, s, etc. Atomic propositions, denoted by lowercase letters, signify statements that can be either true or false.

### Examples of Atomic propositions

Here are examples of atomic propositions:

- p: $5 + 3 = 8$. It is an atomic proposition because it represents a true fact.

- q: Bananas are yellow. It is an atomic proposition because it represents a true fact.

- r: Sun rises in the west. It is an atomic proposition because it represents a false fact.

- s: Moon is black. It is an atomic proposition because it represents a true fact.

## Compound Propositions

Compound propositions are those formed by combining one or more atomic propositions using connectives. If a proposition contains connectives, it is termed a compound proposition. These propositions are constructed by combining simple and atomic propositions with the aid of parentheses and logical connectives. Compound propositions are denoted by capital letters such as P, Q, R, S, etc.

# Examples of Compound Propositions

**Examples of compound propositions include:**

- Today is sunny and I will go to school.

- Oranges are orange, and bananas are yellow.

- The sun sets in the west, and the sun rises in the east.

- John is an engineer, and he works for XYZ Company.

# Statements that are not Propositions

Here are categories of statements that are not propositions:

**Command:** A statement functions as a command when it directs one person to take action. These statements may or may not begin with an imperative (bossy) verb.

**Questions:** A statement serves as a question when it seeks information or clarification from another person. It prompts a response.

**Exclamation:** An exclamation expresses strong emotion. It conveys enthusiasm, surprise, or other intense feelings. Typically, an exclamation mark is used to punctuate such statements.

**Inconsistent:** A statement is inconsistent if it contains a contradiction or self-refuting claim. It implies behavior that is not consistently followed or upheld.

**Predicate:** A predicate statement contains variables and describes a relationship between them. When specific values are assigned to the variables, it transforms into a statement with a definite truth value.

1. In this illustration, we present statements that do not qualify as propositions. They are as follows:

2. "Catch the ball." This statement functions as a command, as it directs one person to retrieve the ball.

3. "Do you have any problem with Harry?" This statement serves as a question, as it seeks clarification from another person regarding Harry.

4. "What an awesome day!" This statement expresses exclamation, as it conveys strong emotion, evident from the exclamation mark.

5. "I always tell lies." This statement is inconsistent, as it contradicts itself by claiming a habitual behavior that cannot be consistently maintained.

6. "P(x) = a - 5 = 10." This statement is a predicate, as it includes a variable, x, with an assigned value.

## Examples of Propositions

Here we present various examples of propositions, each explained as follows:

- "Lucknow is the capital of Uttar Pradesh." This is a true proposition.

- "2024 will be a leap year." This is a true proposition.

- "Mangoes are black." This is a false proposition.

- "P(x) = x + 2 = 4." This is not a proposition because it is a predicate.

- "P(5): 4 + 8 = 10." This is a false proposition because 4 + 8 does not equal 10.

- "Bananas are green." This is a false proposition.

- "Oranges are purple." This is a false proposition.

- "The sum of three and five is eight." This is a true proposition.

- "X is less than 4." This is not a proposition; it is a predicate because the variable x is not assigned a specific value.

- "Go to college." This is not a proposition; it is a command because one person is instructing someone to take action.

- "Are you sure?" This is not a proposition; it is a question because it seeks information from another person.

- "Wow! She is very beautiful." This is not a proposition; it is an exclamation expressing strong emotion, denoted by the exclamation mark.

- "Delhi is not part of India." This is a false proposition.

- "I am a good student." This is a proposition that can be either true or false.

- "I always tell a lie." This is not a proposition; it is inconsistent because one cannot consistently tell a lie.

- "This statement is true." This is a true proposition.

- "This statement is false." This is not a proposition; it is inconsistent.

- "Water is running." This is a proposition that can be either true or false.

- "The sun will rise tomorrow." This is a proposition that can be either true or false.

- "Don't touch my cheeks." This is not a proposition; it is a command instructing someone not to take action.

- "Don't go there." This is not a proposition; it is a command instructing someone not to take action.

## Limitations of Propositional Logic

There are several limitations of propositional logic, outlined below:

1.  Propositional logic does not enable the representation of relations such as "some," "all," or "none." For instance:

    - "All girls in my class are smart and intelligent."

    - "Some mangoes are sweet and tart."

2.  Propositional logic has limited expressive power.

3.  The statements in propositional logic cannot be represented in terms of their properties or logical relationships.

## Important Points of Propositional logic

Here are some important points related to propositional logic:

- Propositional logic is also referred to as Boolean logic because it operates on the basis of binary values, typically represented as 1 and O.

- In propositional logic, logical expressions are symbolically represented using variables such as P, Q, R, X, Y, Z, etc.

- Propositional logic evaluates statements as either true or false, but not both simultaneously.

- It incorporates logical connectives or operators to establish relationships between propositions.

- The fundamental components of propositional logic are propositions (statements) and logical connectives.

- A propositional formula that always evaluates to true is termed a tautology or valid sentence.

- A propositional formula that always evaluates to false is termed a contradiction.

- If a propositional formula can be true or false depending on the values assigned to its variables, it is termed consistent or a contingency.

## 5.2. Logical Connectives

Logical connectivity refers to the operators employed to link one or more propositions or elements of predicate logic. Depending on the input logic and the connectivity utilized to link the propositions, we derive the resultant logic. Propositional logic encompasses five fundamental connectives, outlined as follows:

1. Negation
2. Conjunction
3. Disjunction
4. Conditional
5. Bi-conditional

The terms, connective words, and symbols used in Propositional logic are outlined as follows:

| Name of Connective | Connective Word | Symbol |
| --- | --- | --- |
| Negation | Not | $\rceil$ or ~ or ' or - |
| Conjunction | And | $\wedge$ |
| Disjunction | Or | $\vee$ |
| Conditional | If-then | $\rightarrow$ |
| Bi-conditional | If and only if | $\leftrightarrow$ |

Now, let's examine each of these connectives individually, as listed below:

## Negation

The symbol ~ denotes negation. For a proposition p, its negation, denoted as ~p, exhibits the following properties:

- When p is true, ~p is false.
- When p is false, ~p is true.

## Truth table

Below is the truth table for negation:

| p | ~p |
| --- | --- |
| T | F |
| F | T |

## Example

An illustration demonstrating negation is provided below:

- Suppose p: "I am a good student."
- Then the negation of p will be:
- ¬p: «I am not a good student.»

## Conjunction:

Conjunction is denoted by the symbol $\wedge$, describes the logical operation where two propositions, p and q, are combined.

The conjunction of p and q results in a proposition with the following properties:

- When both p and q are true, their conjunction is true.
- When either p or q (or both) is false, their conjunction is false.

The truth table for conjunction is illustrated below:

| P | q | p ∧ q |
|---|---|-------|
| T | T | T |
| T | F | F |
| F | T | F |
| F | F | F |

## Example

An example illustrating conjunction is as follows:

Given two propositions, p and q, where

1. p: I am a good student.
2. q: I like my math teacher.

Then the conjunction of p and q will be:

p∧q: I am a good student, and I like my math teacher.

# Disjunction

Disjunction, denoted by the symbol ∨, applies to two propositions, p and q, resulting in a new proposition with the following characteristics:

- If both p and q are false, their disjunction is false.
- If either p or q, or both, are true, their disjunction is true.

# Truth table

Below is the truth table illustrating disjunction:

| P | q | p ∨ q |
|---|---|-------|
| T | T | T |
| T | F | T |
| F | T | T |
| F | F | F |

**Example**

Here's an example illustrating disjunction:

Suppose we have two propositions, p and q:

1. p: "I am a good student."

2. q: "I like my math teacher."

The disjunction of p and q would be:

p ∨ q: "I am a good student or I like my math teacher."

## Conditional

The conditional proposition, also referred to as implication, is denoted by the symbol →. When considering two propositions, p and q, the conditional statement formed from them is another proposition with the following properties:

- Any proposition in the form "if p then q" is categorized as an implication or conditional proposition.

- The implication is true when either p is false or both p and q are true.

- It is false when p is true and q is false.

## Truth table

Below is the truth table illustrating implication:

| P | q | p → q |
|---|---|-------|
| T | T | T |
| T | F | F |
| F | T | T |
| F | F | T |

## Example

Here's an example illustrating the implication proposition:

1. If x equals y and y equals z, then x equals c.
2. If I study very hard, then I will get good marks in the exam.

## Bi-conditional

The bi-conditional proposition, also referred to as bi-implication, is denoted by the symbol ↔. When considering two propositions, p and q, the bi-conditional statement formed from them is another proposition with the following properties:

- Any proposition in the form "p if and only if q" is categorized as a bi-implication or bi-conditional proposition.

- The bi-conditional is true when both p and q are true, or when both p and q are false.

- In all other cases, the bi-conditional is false.

## Truth table

Below is the truth table illustrating bi-implication:

| P | Q | p ↔ q |
|---|---|---|
| T | T | T |
| T | F | F |
| F | T | F |
| F | F | T |

**Example**

Here's an example illustrating the bi-implication proposition:

1. I will go to the beach if and only if it is sunny.

2. I will get good marks if and only if I study hard.

**Note 1:**

- Each logical connective must have a certain priority assigned to it.

- During problem-solving, the sequence of this priority becomes significant.

- The decreasing order of priority is depicted in the following image.

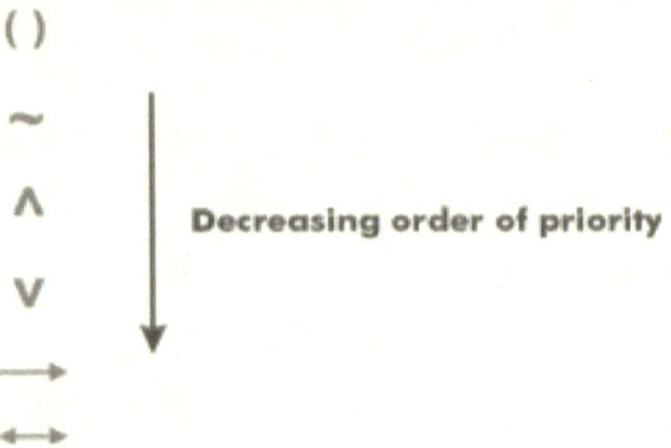


**Note 2:**

- The negation, disjunction, conjunction, and bi-implication or bi-conditional operators encompass the commutative and associative properties.

## 5.3. WELL FORMED FORMULAS

A Well-Formed Formula (WFF) is an expression comprised of variables (capital letters), parentheses, and connective symbols. In this context, an expression is essentially a composition of operands and operators, with the operands being the variables and the operators being the connective symbols.

Below are the possible Connective Symbols

1. ¬ **(Negation)**

2. ∧ **(Conjunction)**

3. ∨ **(Disjunction)**

4. ⇒ **(Rightwards Arrow)**

5. ⇔ **(Left-Right Arrow)**

### Statement Formulas

1.  Statements lacking any connectives are termed as Atomic or Simple statements, and these statements alone qualify as WFFs.

    *For example,*

    P, Q, R, etc.

2.  Statements incorporating one or more elementary statements are referred to as Molecular or Composite statements.

    For instance, if P and Q represent two simple statements, then several composite statements adhering to WFF standards can be constructed, such as:

$\rightarrow \neg P$

$\rightarrow \neg Q$

$\rightarrow (P \vee Q)$

$\rightarrow (P \wedge Q)$

$\rightarrow (\neg P \vee Q)$

$\rightarrow ((P \vee Q) \wedge Q)$

$\rightarrow (P \Rightarrow Q)$

$\rightarrow (P \Leftrightarrow Q)$

$\rightarrow \neg(P \vee Q)$

$\rightarrow \neg(\neg P \vee \neg Q)$

## Rules of the Well-Formed Formulas

1. An isolated statement variable constitutes a Well-Formed Formula (WFF). For instance, statements like P, ~ P, Q, ~Q are WFFs by themselves.

2. If 'P' is a WFF, then ~P is also a formula.

3. If P & Q are WFFs, then (P∨Q), (P∧Q), (P⇒Q), (P⇔Q), etc., are likewise WFFs.

4. A sequence of symbols involving statement variables, connectives, and parentheses qualifies as a Well-Formed Formula if and only if it can be derived through a finite number of applications of rules 1, 2, and 3.

## Example Of Well Formed Formulas:

| WFF | Explanation |
| --- | --- |
| $\neg\neg P$ | By **Rule 1** each Statement by itself is a WFF, $\neg P$ is a WFF, and let $\neg P = Q$. So $\neg Q$ will also be a WFF. |
| $((P\Rightarrow Q)\Rightarrow Q)$ | By **Rule 3** joining '$(P\Rightarrow Q)$' and 'Q' with connective symbol '$\Rightarrow$'. |
| $(\neg Q \wedge P)$ | By **Rule 3** joining '$\neg Q$' and 'P' with connective symbol '$\wedge$'. |
| $((\neg P \vee Q)\wedge\neg\neg Q)$ | By **Rule 3** joining '$(\neg P\vee Q)$' and '$\neg\neg Q$' with connective symbol '$\wedge$'. |
| $\neg((\neg P\vee Q)\wedge\neg\neg Q)$ | By **Rule 3** joining '$(\neg P\vee Q)$' and '$\neg\neg Q$' with connective symbol '$\wedge$' and then using Rule 2. |

**Below are examples that may resemble WFFs but are not considered Well-Formed Formulas:**

1. (P): While 'P' alone is considered a WFF by Rule 1, enclosing it in parentheses is not recognized as a WFF by any rule.

2. $\neg P \wedge Q$: This can be interpreted as either $(\neg P\wedge Q)$ or $\neg(P\wedge Q)$, leading to ambiguity, hence it is not considered a WFF. Parentheses are necessary in Composite Statements.

3. $((P \Rightarrow Q))$: While $(P\Rightarrow Q)$ is a WFF, considering the outer parentheses leaves us with (A) if we let $(P\Rightarrow Q) = A$, which is not a valid WFF. Parentheses are crucial in such cases.

4. $(P \Rightarrow\Rightarrow Q)$: Placing a connective symbol right after another connective symbol is not valid for a WFF.

5. $((P \wedge Q) \wedge)Q)$: Having a conjunction operator after $(P \wedge Q)$ is invalid.

6. $((P \wedge Q) \wedge PQ)$: Invalid placement of variables $(PQ)$.

7. $(P \vee Q) \Rightarrow (\wedge Q)$: With the Conjunction component, only one variable 'Q' is present. At least two variables are required to form an operation inside parentheses.

**Example: Check whether the following formulas are well formed formulas or not.**

i)    $\neg (P \wedge Q)$                           - Well Formed Formula

ii)   $\neg (P \vee Q)$                           - Well Formed Formula

iii)  $\neg P \wedge Q$                            - Not Well Formed Formula

iv)   $(P \Rightarrow Q) \Rightarrow (\wedge Q)$                     - Not Well Formed Formula

v)    $(P \Rightarrow (P \vee Q))$                      - Well Formed Formula

vi)   $(P \Rightarrow (Q \Rightarrow R))$                    - Well Formed Formula

vii)  $(P \Rightarrow Q$                            - Not Well Formed Formula

viii) $(P \wedge Q) \Rightarrow Q)$                     - Not Well Formed Formula

ix)   $(((P \Rightarrow Q) \wedge (Q \Rightarrow R)) \Leftrightarrow (P \Rightarrow R))$ - Well Formed Formula

## 5.4. TAUTOLOGIES

A tautology refers to a compound statement that remains true regardless of the truth values of its individual components. This term originates from the Greek words "tauto," meaning "same," and "logy," referring to "logic." Conditional words like if, then, and, or, not, and if and only if are utilized to construct compound statements, positioning them between two simple statements. For instance, given statements A and B, the expression $(A \Rightarrow B) \vee (B \Rightarrow A)$ exemplifies a tautology.

Examples of tautologies are straightforward:

- "Jack is a good boy, or Jack is not a good boy."
- "Either he will choose to study biology, or he will not choose to study biology."
- "A number can be even, or a number cannot be even."

Tautologies can be represented as compound statements consistently yielding a true value, unaffected by the individual truth values of its components. They can be readily translated from ordinary language to mathematical expressions using logical symbols. For instance, the statement "My uncle gives me 100 rupees or my uncle will not give me 100 rupees" can be interpreted as follows:

1. Let P represent "My uncle gives me 100 rupees."
2. Let ~P represent "My uncle will not give me 100 rupees" (the opposite of statement P).

The logical operator "OR" (denoted by the symbol "$\vee$") combines these two statements, leading to:

$$P \vee {\sim}P$$

To ascertain the validity of this expression, we examine two cases:

**Case 1:** My uncle gives me 100 rupees. In this scenario, the first statement yields true, while the second yields false. Through the 'OR' operator, these statements are connected, resulting in a truth statement.

**Case 2:** My uncle will not give me 100 rupees. Here, the first statement yields false, while the second yields true, still resulting in a true statement.

To further analyze this, we utilize a truth table, outlined below:

| P = My uncle gives me 100 rupees | ~P = My uncle will not give me 100 rupees | P ∨ ~P (My uncle gives me 100 rupees or My uncle will not give me 100 rupees) |
| --- | --- | --- |
| T | F | T |
| F | T | T |

Therefore, as observed from the final column of the table above, it is evident that the statement consistently evaluates to true across all values. Thus, we can conclude that the given statement is indeed a tautology.

## Tautology Logic Symbols

Compound statements can be depicted using tautologies employing various logical symbols. The distinct types of symbols, their significance, and their representation in discrete mathematics are elaborated as follows:

| Symbols | Meaning | Representation |
| --- | --- | --- |
| ∨ | OR | A ∨ B |
| ¬ | Negation | ¬A |
| ~ | NOT | ~ A |
| ∧ | AND | A ∧ B |
| = | Is equivalent to | A = B |
| → | If-then or Implies | A → B |
| ⇔ | If and only if | A ⇔ B |

## Examples of Tautology

There are various examples of tautology, which are described as follows:

**Example 1:** In this example, we have to determine whether the statement ~h ⇒ h is a tautology.

**Solution:**

There is a statement h. Truth value of this statement can be written in the following form:

| **H** | **~h** | **~h ⇒ h** |
| --- | --- | --- |
| T | F | T |
| F | T | F |

With the help of above table, we can see that the truth value of ~h ⇒ h is {T, F}. That's why this statement is not a tautology.

**Example 2:** In this example, we have to determine whether the statement p ⇒ (p ∨ q) is a tautology.

**Solution:**

There are two statements, p, and q. Truth value of these statements can be written in the following form:

| p | q | p ∨ q | p ⇒ (p ∨ q) |
| --- | --- | --- | --- |
| T | T | T | T |
| T | F | T | T |
| F | T | T | T |
| F | F | F | T |

With the help of above table, we can see that the truth value of p ⇒ (p ∨ q) is true for all the individual statements. That's why this statement is a tautology.

**Example 3:** In this example, we have to determine whether the statement ~A ∧ B ⇒ ~(A ∨ B) is a tautology.

**Solution:**

There are two statements, A, and B. Truth value of these statements can be written in the following form:

| A | ~A | B | ~A∧B | A∨B | ~(A∨B) | ~ A ∧ B ⇒ ~(A ∨ B) |
|---|----|---|------|-----|--------|----------------------|
| T | F | T | F | T | F | T |
| T | F | F | F | T | F | T |
| F | T | T | T | T | F | F |
| F | T | F | F | F | T | T |

With the help of above table, we can see that the truth value of ~A ∧ B ⇒ ~(A ∨ B) is not true for all the individual statements. That's why this statement is not a tautology.

## Tautology and Contradiction

In the preceding explanation, we've become acquainted with the term "tautology." Its counterpart, known as the contradiction, emerges when two statements are amalgamated to form a compound statement using logical operations, yielding a false outcome. The contradiction is synonymous with fallacy. For instance, consider two statements, A and B. If (A ⇒ B) ∨ (B ⇒ A) represents a tautology, then ~(A ⇒ B) ∨ (B ⇒ A) would be deemed a contradiction or fallacy.

| A | B | A⇒B | B⇒A | Tautology = $(A \Rightarrow B) \lor (B \Rightarrow A)$ | Contradiction = $\sim(A \Rightarrow B) \lor (B \Rightarrow A)$ |
|---|---|---|---|---|---|
| T | T | T | T | T | F |
| T | F | F | T | T | F |
| F | T | T | F | T | F |
| F | F | T | T | T | F |

**Example 1:** Prove that $[\,(\,P \Rightarrow Q\,) \land (\,Q \Rightarrow R\,)\,] \Rightarrow [\,P \Rightarrow R\,]$ is a tautology.

| P | Q | R | P⇒Q a | Q⇒R b | a∧b | P⇒R c | a∧b⇒c |
|---|---|---|---|---|---|---|---|
| T | T | T | T | T | T | T | T |
| T | T | F | T | F | F | F | T |
| T | F | T | F | T | F | T | T |
| T | F | F | F | T | F | F | T |
| F | T | T | T | T | T | T | T |
| F | T | F | T | F | F | T | T |
| F | F | T | T | T | T | T | T |
| F | F | F | T | T | T | T | T |

With the help of above table, we can see that the truth value of $[\,(\,P \Rightarrow Q\,) \land (\,Q \Rightarrow R\,)\,] \Rightarrow [\,P \Rightarrow R\,]$ is true for all the individual statements. That›s why this statement is a tautology.

**Example 2:** Prove that $\sim[\,(\,P \vee Q\,) \wedge \sim P\,] \Rightarrow Q$ is a contradiction.

| P | Q | ~P | P ∨ Q | b ∧ a | (b ∧ a) ⇒ Q | ~c |
|---|---|----|-------|-------|-------------|----|
|   |   | a | b |  | c |  |
| T | T | F | T | F | T | F |
| T | F | F | T | F | T | F |
| F | T | T | T | T | T | F |
| F | F | T | F | F | T | F |

With the help of above table, we can see that the truth value of $\sim[\,(\,P \vee Q\,) \wedge \sim P\,] \Rightarrow Q$ is false for all the individual statements. That's why this statement is a contradiction.

## 5.5. EQUIVALENCES

Consider two compound statements, X and Y, which are deemed logically equivalent if and only if their truth tables yield identical truth values in all columns. This logical equivalence is denoted by the symbol = or ⇔, thus X = Y or X ⇔ Y signifies the equivalence of these statements.

Through the definition of logical equivalence, we establish that if compound statements X and Y are indeed logically equivalent, then X ⇔ Y must constitute a tautology.

### Laws of Logical Equivalence

Within these laws, we utilize the symbols 'AND' and 'OR' to elucidate logical equivalence. 'AND' is represented by the ∧ symbol, while 'OR' is represented by the ∨ symbol. Various laws of logical equivalence are delineated as follows:

## Idempotent Law

The idempotent law focuses on a single statement. According to this law, combining two identical statements with the symbols $\wedge$ (AND) and $\vee$ (OR) yields the original statement. Let's consider a compound statement P. The idempotent law is represented as follows:

1. $P \vee P = P$
2. $P \wedge P = P$

The truth table for this law is described as follows:

| P | P | $P \vee P$ | $P \wedge P$ |
|---|---|---|---|
| T | T | T | T |
| F | F | F | F |

This table contains the same truth values in the columns of P, $P \vee P$ and $P \wedge P$.

Hence we can say that $P \vee P = P$ and $P \wedge P = P$.

## Commutative Laws

Two statements are employed to illustrate the commutative law. As per this law, when two statements are combined with the symbols $\wedge$ (AND) or $\vee$ (OR), the resultant statement remains unchanged even if we alter the positions of the statements. Consider two statements, P and Q. The compound statement formed by these statements yields a false proposition only when both P and Q are false. Otherwise, it evaluates to true. The commutative law is represented as follows:

1. $P \vee Q = Q \vee P$
2. $P \wedge Q = Q \wedge P$

The truth table for these notations is described as follows:

| P | Q | P $\vee$ Q | Q $\vee$ P |
|---|---|---|---|
| T | T | T | T |
| T | F | T | T |
| F | T | T | T |
| F | F | F | F |

This table contains the same truth values in the columns of P $\vee$ Q and Q $\vee$ P.

Hence we can say that P $\vee$ Q = Q $\vee$ P.

Same as we can prove P $\wedge$ Q = Q $\wedge$ P.

## Associative Law

Three statements are utilized to demonstrate the associative law. In accordance with this law, when three statements are combined using brackets and the symbols $\wedge$ (AND) or $\vee$ (OR), the resulting statement remains unchanged even if we rearrange the order of the brackets. This signifies that the law remains consistent regardless of grouping or association. Consider three statements, P, Q, and R. The compound statement formed by these statements yields a false proposition only when all three P, Q, and R are false. Otherwise, it evaluates to true.

The associative law is represented as follows:

1.  P $\vee$ (Q $\vee$ R) = (P $\vee$ Q) $\vee$ R
2.  P $\wedge$ (Q $\wedge$ R) = (P $\wedge$ Q) $\wedge$ R

The truth table for these notations is described as follows:

| P | Q | R | P ∨ Q | Q ∨ R | (P ∨ Q) ∨ R | P ∨ (Q ∨ R) |
|---|---|---|---|---|---|---|
| T | T | T | T | T | T | T |
| T | T | F | T | T | T | T |
| T | F | T | T | T | T | T |
| T | F | F | T | F | T | T |
| F | T | T | T | T | T | T |
| F | T | F | T | T | T | T |
| F | F | T | F | T | T | T |
| F | F | F | F | F | F | F |

This table contains the same truth values in the columns of P ∨ (Q ∨ R) and (P ∨ Q) ∨ R.

Hence we can say that P ∨ (Q ∨ R) = (P ∨ Q) ∨ R.

Same as we can prove P ∧ (Q ∧ R) = (P ∧ Q) ∧ R

## Distributive Law

Three statements are employed to illustrate the distributive law. As per this law, if a statement joined by the ∨ (OR) symbol is combined with two other statements linked by the ∧ (AND) symbol, the resulting statement remains unchanged even if we separately combine the statements with the ∨ (OR) symbol and then combine the joined statements with ∧ (AND). Consider three statements, P, Q, and R. The distributive law is represented as follows:

P ∨ (Q ∧ R) = (P ∨ Q) ∧ (P ∨ R)

P ∧ (Q ∨ R) = (P ∧ Q) ∨ (P ∧ R)

The truth table for these notations is described as follows:

| P | Q | R | Q∧R | P∨(Q∧R) | P∨Q | P∨R | (P ∨ Q) ∧ (P ∨ R) |
|---|---|---|-----|---------|-----|-----|-------------------|
| T | T | T | T | T | T | T | T |
| T | T | F | F | T | T | T | T |
| T | F | T | F | T | T | T | T |
| T | F | F | F | T | T | T | T |
| F | T | T | T | T | T | T | T |
| F | T | F | F | F | T | F | F |
| F | F | T | F | F | F | T | F |
| F | F | F | F | F | F | F | F |

This table contains the same truth values in the columns of $P \vee (Q \wedge R)$ and $(P \vee Q) \wedge (P \vee R)$.

Hence we can say that $P \vee (Q \wedge R) = (P \vee Q) \wedge (P \vee R)$

Same as we can prove $P \wedge (Q \vee R) = (P \wedge Q) \vee (P \wedge R)$

## Identity Law

The identity law is demonstrated using a single statement. According to this law, if we combine a statement with a True value using the symbol $\vee$ (OR), it will yield a True value. Conversely, if we combine a statement with a False value using the symbol $\wedge$ (AND), it will yield the statement itself. The same principle applies when using the opposite symbols. This means that combining a statement with a True value using the symbol $\wedge$ (AND) results in the statement itself, and combining a statement with a False value using the symbol $\vee$ (OR) yields a False value. Let's consider a compound statement P, a True value T, and a False value F. The identity law is represented as follows:

1.  $P \vee T = T$ and $P \vee F = P$

2.  $P \wedge T = P$ and $P \wedge F = F$

The truth table for these notations is described as follows:

| P | T | F | P ∨ T | P ∨ F |
|---|---|---|-------|-------|
| T | T | F | T | T |
| F | T | F | T | F |

This table contains the same truth values in the columns of $P \vee T$ and T. Hence, we can say that $P \vee T = T$. Similarly, this table also contains the same truth values in the columns of $P \vee F$ and P. Hence we can say that $P \vee F = P$.

Same as we can prove $P \wedge T = P$ and $P \wedge F = F$

## Complement Law

The complement law involves a single statement. According to this law, if we combine a statement with its complement using the symbol $\vee$ (OR), it will yield a True value. Conversely, combining these statements with the symbol $\wedge$ (AND) will yield a False value. When a True value is negated, it results in a False value, and when a False value is negated, it yields a True value.

The following notation represents the complement law:

1.  $P \vee \neg P = T$ and $P \wedge \neg P = F$

2.  $\neg T = F$ and $\neg F = T$

The truth table for these notations is described as follows:

| P | ¬P | T | ¬T | F | ¬F | P ∨ ¬P | P ∧ ¬P |
|---|----|---|----|---|----|--------|--------|
| T | F | T | F | F | T | T | F |
| F | T | T | F | F | T | T | F |

The truth values in the columns of P ∨ ¬P and T are identical in this table. Therefore, we can conclude that P ∨ ¬P = T. Similarly, the truth values in the columns of P ∧ ¬P and F are the same. Thus, we can assert that P ∧ ¬P = F.

Likewise, the truth values in the columns of ¬T and F match in this table. Thus, we can deduce that ¬T = F. Similarly, the truth values in the columns of ¬F and T are identical. Therefore, we can state that ¬F = T.

## Double Negation Law or Involution Law

The double negation law is demonstrated using a single statement. According to this law, if we negate a negated statement, the resulting statement will be the original statement itself. Let's consider a statement P and its negation ¬P.

The following notation represents the Double negation law:

1.  ¬(¬P) = P

The truth table for these notations is described as follows:

| P | ¬P | ¬(¬P) |
|---|----|-------|
| T | F | T |
| F | T | F |

This table contains the same truth values in the columns of ¬(¬P) and P. Hence we can say that ¬(¬P) = P.

## De Morgan's Law

De Morgan's law is demonstrated using two statements. According to this law, if we combine two statements with the symbol ∧ (AND) and then negate the combined statements, the resulting statement remains the same even if we separately combine the negations of both statements with the symbol $\vee$ (OR). Let's consider two compound statements, P and Q.

The following notation represents De Morgan's Law:

1. $\neg(P \wedge Q) = \neg P \vee \neg Q$

2. $\neg(P \vee Q) = \neg P \wedge \neg Q$

The truth table for these notations is described as follows:

| P | Q | ¬P | ¬Q | P∧Q | ¬(P∧Q) | ¬P∨¬Q |
|---|---|----|----|-----|--------|-------|
| T | T | F  | F  | T   | F      | F     |
| T | F | F  | T  | F   | T      | T     |
| F | T | T  | F  | F   | T      | T     |
| F | F | T  | T  | F   | T      | T     |

The truth values in the columns of $\neg(P \wedge Q)$ and $\neg P \vee \neg Q$ are identical in this table. Therefore, we can conclude that $\neg(P \wedge Q) = \neg P \vee \neg Q$. Similarly, we can demonstrate that $\neg(P \vee Q) = \neg P \wedge \neg Q$.

## Absorption Law

The absorption law is demonstrated using two statements. According to this law, if we combine a statement P with the $\vee$ (OR) symbol with the same statement P and another statement Q, joined by the $\wedge$ (AND) symbol, the resulting statement will be the initial statement P. The same outcome is obtained when the symbols are interchanged. Let's consider two compound statements, P and Q.

The following notation represents the Absorption Law:

1. $P \vee (P \wedge Q) = P$

2. $P \wedge (P \vee Q) = P$

The truth table for these notations is described as follows:

| P | Q | P∧Q | P∨Q | P∨(P∧Q) | P∧(P∨Q) |
|---|---|-----|-----|---------|---------|
| T | T | T | T | T | T |
| T | F | F | T | T | T |
| F | T | F | T | F | F |
| F | F | F | F | F | F |

The truth values in the columns of P ∨ (P ∧ Q) and P are identical in this table. Therefore, we can conclude that P ∨ (P ∧ Q) = P. Similarly, the table also contains the same truth values in the columns of P ∧ (P ∨ Q) and P. Hence, we can assert that P ∧ (P ∨ Q) = P.

## Examples of Logical Equivalence

There are various examples of logical equivalence. Some of them are described as follows:

**Example 1:** In this example, we will establish the equivalence property for a statement, which is described as follows:

$$p \rightarrow q = \neg p \lor q$$

**Solution:**

We will prove this with the help of a truth table, which is described as follows:

| P | Q | ¬P | P→Q | ¬P∨Q |
|---|---|----|-----|------|
| T | T | F | T | T |
| T | F | F | F | F |
| F | T | T | T | T |
| F | F | T | T | T |

This table contains the same truth values in the columns of $P \rightarrow Q$ and $\neg P \vee Q$. Hence we can say that $P \rightarrow Q = \neg P \vee Q$.

**Example 2:** In this example, we will establish the equivalence property for a statement, which is described as follows:

$$P \leftrightarrow Q = (P \rightarrow Q) \wedge (Q \rightarrow P)$$

**Solution:**

| P | Q | $P \rightarrow Q$ | $Q \rightarrow P$ | $P \leftrightarrow Q$ | $(P \rightarrow Q) \wedge (Q \rightarrow P)$ |
|---|---|---|---|---|---|
| T | T | T | T | T | T |
| T | F | F | T | F | F |
| F | T | T | F | F | F |
| F | F | T | T | T | T |

This table contains the same truth values in the columns of $P \leftrightarrow Q$ and $(P \rightarrow Q) \wedge (Q \rightarrow P)$. Hence we can say that $P \leftrightarrow Q = (P \rightarrow Q) \wedge (Q \rightarrow P)$.

**Example 3:** In this example, we will use the equivalent property to prove the following statement:

$$P \leftrightarrow Q = (P \wedge Q) \vee (\neg P \wedge \neg Q)$$

**Solution:**

To prove this, we will use some of the above-described laws and from this law we have:

$$P \leftrightarrow Q = (\neg P \vee Q) \wedge (\neg Q \vee P) \qquad (1)$$

Now we will use the Commutative law in the above equation and get the following:

$$= (\neg P \vee Q) \wedge (P \vee \neg Q)$$

Now we will use the Distributive law in this equation and get the following:

$$= (\neg P \wedge (P \vee \neg Q)) \vee (Q \wedge (P \vee \neg Q))$$

Now we will use Distributive law in this equation and get the following:

$$= (\neg P \wedge P) \vee (\neg P \wedge \neg Q) \vee (Q \wedge P) \vee (Q \wedge \neg Q)$$

Now we will use the complement law in this equation and get the following:

$$= F \vee (\neg P \wedge \neg Q) \vee (Q \wedge P) \vee F$$

Now we will use the identity law and get the following:

$$= (\neg P \wedge \neg Q) \vee (Q \wedge P)$$

Now we will use the Commutative law in this equation and get the following:

$$= (P \wedge Q) \vee (\neg P \wedge \neg Q)$$

Finally, equation (1) becomes the following:

$$P \leftrightarrow Q = (P \wedge Q) \vee (\neg P \wedge \neg Q)$$

Finally, we can say that the equation (1) becomes $P \leftrightarrow Q = (P \wedge Q) \vee (\neg P \wedge \neg Q)$

## Laws of Logic

**1    Idempotent Laws**

   a)   $P \vee P = P$

   b)   $P \wedge P = P$

**2 Commutative Laws**

a)   $P \vee Q = Q \vee P$

b)   $P \wedge Q = Q \wedge P$

**3 Identity Laws**

a)   $P \wedge T = P$

b)   $P \vee F = P$

**4 Domination Laws**

a)   $P \vee T = T$

b)   $P \wedge F = F$

**5 Associative Laws**

a)   $(P \vee Q) \vee R = P \vee (Q \vee R)$

b)   $(P \wedge Q) \wedge R = P \wedge (Q \wedge R)$

**6 Distributive Laws**

a)   $P \vee (Q \wedge R) = (P \vee Q) \wedge (P \vee R)$

b)   $P \wedge (Q \vee R) = (P \wedge Q) \vee (P \wedge R)$

**7 Complement Laws**

a)   $P \vee \sim P = T$

b)   $P \wedge \sim P = F$

**8 Double Negation**

a)   $\sim(\sim P) = P$

**9 DeMorgan's Laws**

a)   $\sim (P \wedge Q) = \sim P \vee \sim Q$

b)   $\sim (P \vee Q) = \sim P \wedge \sim Q$

**10  Absorption Laws**

    a)   $P \vee (P \wedge Q) = P$

    b)   $P \wedge (P \vee Q) = P$

**11**  a) $P \Rightarrow Q = \sim P \vee Q$

    b) $\sim (P \Rightarrow Q) = \sim (\sim P \vee Q) = P \wedge \sim Q$

**12**  a) $P \Leftrightarrow Q = (\sim P \vee Q) \wedge (\sim Q \vee P)$

    b) $\sim [P \Leftrightarrow Q] = \sim [(\sim P \vee Q) \wedge (\sim Q \vee P)] = (P \wedge \sim Q) \vee (Q \wedge \sim P)$

## 5.6. INFERENCE THEORY

The main aim of logic is to provide rules of inference to infer a conclusion from certain premises. The theory associated with rules of inference is known as "Inference Theory".

### Valid Argument or Valid Conclusion

If a conclusion is derived from a set of premises by using the accepted rules of reasoning, then such a process of derivation is called a "Deduction" or "a formal proof" and the argument or conclusion is called a "Valid argument" or "Valid Conclusion".

**Note:**

Premises are also called as assumptions (or) axioms (or) hypotheses.

**Methods used to determine the conclusion**

The following methods used to determine whether the conclusion logically follows from the given premises.

    i)   Using Truth Table method

    ii)  Without Using Truth Table method

## Using Truth Table Method

Let P1, P2, ... Pn be the variables appearing in the premises H1, H2, ... Hm and the conclusion C.

a) Look for the rows in which all H1, H2, ... Hm have the value True, if for every such row, C also has the value True, then H1  H2  ... Hm ➜ C holds.

b) Look for the rows in which C has the value False, if in every such row at least one of the values of H1, H2, ... Hm is False then H1  H2  ... Hm ➜ C holds.

**Example:**

Determine if the conclusion C follows logically from the given premises.

i)   H1: $P \rightarrow Q$          H2: P          C: Q

| | H2 | | C | | H1 |
|---|---|---|---|---|---|
| | P | | Q | | $P \rightarrow Q$ |
| | T | | T | | T |
| | T | | F | | F |
| | F | | T | | T |
| | F | | F | | T |

From the above table, when H1 and H2 are True, C is also True. Hence the given premises produces a valid conclusion.

ii)  H1: P → Q          H2: Q          C: P

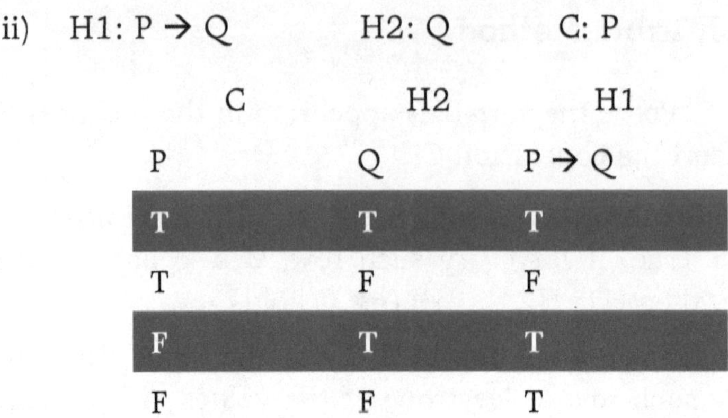

|  C  |  H2  |  H1  |
| --- | --- | --- |
|  P  |  Q  |  P → Q  |
|  T  |  T  |  T  |
|  T  |  F  |  F  |
|  F  |  T  |  T  |
|  F  |  F  |  T  |

In the above table, when H1 and H2 are True, there are two instances and C is True for one instance and it is False for the next instance. Hence for the given premises there is no valid conclusion.

iii)  H1: P → Q          H2: P          C: Q

|  |  |  C  |  H2  |  H1  |
| --- | --- | --- | --- | --- |
|  P  |  Q  |  |  ¬P  |  P → Q  |
|  T  |  T  |  |  F  |  T  |
|  T  |  F  |  |  F  |  F  |
|  F  |  T  |  |  T  |  T  |
|  F  |  F  |  |  T  |  T  |

In the above table, when H1 and H2 are True, there are two instances and C is True for one instance and it is False for the next instance. Hence for the given premises there is no valid conclusion.

iv)   H1: QH2: P → Q   C: P

| P | Q | C<br>¬P | H1<br>¬Q | H2<br>P → Q |
|---|---|---|---|---|
| T | T | F | F | T |
| T | F | F | T | F |
| F | T | T | F | T |
| F | F | T | T | T |

In the above table, when H1 and H2 are True, C is also True. Hence the given premises produces a valid conclusion.

## Without Using Truth Table Method

The truth table technique becomes tedious when the number of statement variables present in all the formulas representing the premises and the conclusion is large. To over come this disadvantage, we use the other methods without using Truth Table. Here we describe the process of derivation by which one demonstrate that a particular formula is valid of a given set of premises. Here we have three rules of Inference called (i) Rule P (ii) Rule T (iii) Rule CP.

## Rule P

A premise may be introduced at any point in the derivation.

## Rule T

A formula 'S' may be introduced in a derivation, if 'S' is tautologically implied by one or more of the preceding formulas in the derivation.

## Rule CP

If we can derive 'S' from 'R' and a set of premises, then we can derive R → S from the set of premises alone

For applying these rules, the following Implications and Equivalences are used.

## Implications

I9:  P, Q                    →    P Q

I10: ¬P, P ∨ Q               →    Q

I11: P, P → Q                →    Q

I12: ¬Q, P → Q               →    P

I13: P → Q, Q → R            →    P → R

I14: P ∨ Q, P → R, Q → R     →    R

## Equivalences

E1:   ¬ ( ¬ P)               ≡    P

E10:  P ∨ P                  ≡    P

E11:  P ∧ P                  ≡    P

E16:  P → Q                  ≡    ¬ P ∨ Q

E17:  ¬ (P → Q)              ≡    P ∧ ¬ Q

E18:  P → Q                  ≡    ¬ Q → ¬ P

E19:  P → (Q → R)            ≡    (P ∧ Q) → R

E21:  P ⟨⇒⟩ Q                ≡    (P → Q) ∧ (Q → P)

E22:  P ⟨⇒⟩ Q                ≡    (P ∧ Q) ∨ (¬ P ∧ ¬ Q)

## Example 1:

Show that R  (P ∨ Q) is a valid conclusion from the premises

$$P \vee Q, \, Q \rightarrow R, \, P \rightarrow M, \, \neg M$$

**Solution:**

| [1] | (1) | $P \rightarrow M$ | Rule P |
| [2] | (2) | $\neg M$ | Rule P |
| [1, 2] | (3) | $\neg P$ | Rule T, (1), (2) and I12 |
| [4] | (4) | $P \vee Q$ | Rule P |
| [1, 2, 4] | (5) | $Q$ | Rule T, (3), (4) and I10 |
| [6] | (6) | $Q \rightarrow R$ | Rule P |
| [1, 2, 4, 6] | (7) | $R$ | Rule T, (5), (6) and I11 |
| [1, 2, 4, 6] | (8) | $R \wedge (P \vee Q)$ | Rule T, (4), (7) and I9 |

Therefore R  (P $\vee$ Q) is a valid conclusion from the given premises.

**Example 2:**

Using Rule CP, show that R $\rightarrow$ S is a valid conclusion for the give premises P $\rightarrow$ (Q $\rightarrow$ S), $\neg$ R$\vee$ P, Q

**Solution:**

Given premises: P $\rightarrow$ (Q $\rightarrow$ S), $\neg$ R $\vee$ P, Q

Conclusion: R $\rightarrow$ S

R can be taken as additional premise.

We have to show that the additional premise R

and the given existing premises produces conclusion S.

The premises are:  P $\rightarrow$ (Q $\rightarrow$ S), $\neg$ R $\vee$ P, Q, R

| | | | |
|---|---|---|---|
| [1] | (1) | $\neg R \vee P$ | Rule P |
| [1] | (2) | $R \rightarrow P$ | Rule T, (1) and E16 |
| [3] | (3) | $R$ | Rule P |
| [1, 3] | (4) | $P$ | Rule T, (2), (3) and I11 |
| [5] | (5) | $P \rightarrow (Q \rightarrow S)$ | Rule P |
| [1, 3, 5] | (6) | $Q \rightarrow S$ | Rule T, (4), (5) and I11 |
| [7] | (7) | $Q$ | Rule P |
| [1, 3, 5, 7] | (8) | $S$ | Rule T, (6), (7) and I11 |
| [9] | (9) | $R \rightarrow S$ | Rule CP |

# Appendix

# GLOSSARY

| | |
|---|---|
| Abel's Summation Formula | Abel's summation formula is a mathematical technique used to evaluate infinite series by expressing them as a finite sum involving the original series and its partial sums. |
| Absorption Law | In the context of logic, the absorption law states that in a Boolean algebra, the expression $A \vee (A \wedge B)$ is logically equivalent to A. |
| Adjacency List | An adjacency list is a data structure used in graph theory where each vertex is represented by a list of its neighboring vertices. |
| Adjacency Matrix | An adjacency matrix is a square matrix representing a graph where each cell indicates whether there is an edge between the vertices corresponding to the row and column indices. |
| Antisymmetric Relation | An anti symmetric relation is a binary relation on a set where if (a,b) is in the relation and (b,a) is also in the relation, then a must be equal to b. |
| Associative Law | The associative law stipulates that the grouping of logical operations does not alter the truth value, meaning that $(A \wedge B) \wedge C$ is equivalent to $A \wedge (B \wedge C)$ and $(A \vee B) \vee C$ is equivalent to $A \vee (B \vee C)$. |
| Asymmetric Relation | An asymmetric relation is a relation in which if a is related to b, then b is not related to a. |
| Asymptotic Notation | Asymptotic notation is a mathematical notation used to describe the behavior of a function as its input approaches infinity, commonly used in computer science to analyze algorithm efficiency. |

| | |
|---|---|
| Atomic Proposition | An atomic proposition is a basic statement or proposition that cannot be further divided into simpler statements and serves as the building block of logical expressions. |
| Balanced Tree | A balanced tree is a data structure where the heights of the sub trees of any node differ by at most one, minimizing the depth of the tree and optimizing search and retrieval operations. |
| Bi Conditional | A bi conditional is a logical connective that asserts the mutual implication between two statements, meaning both statements are true or both are false. |
| Big Oh | Big O notation is a mathematical notation used to describe the upper bound or worst-case scenario of the growth rate of a function to analyze algorithm efficiency. |
| Big Omega | Big Omega notation is a mathematical notation used to describe the lower bound or best-case scenario of the growth rate of a function to analyze algorithm efficiency. |
| Big Theta | Big Theta notation is a mathematical notation used to describe the tight bound or average-case scenario of the growth rate of a function to analyze algorithm efficiency. |
| Bijection | A bijection is a function between two sets where each element in the domain corresponds to exactly one element in the codomain, and vice versa, establishing a one-to-one and onto relationship. |

| | |
|---|---|
| Binary Tree | A binary tree is a hierarchical data structure composed of nodes, where each node has at most two children nodes, typically referred to as the left child and the right child. |
| Bipartite Graph | A bipartite graph is a graph whose vertices can be divided into two disjoint sets such that no two vertices within the same set are adjacent, while all edges connect vertices from different sets. |
| Bounding Summations | Bounding summations involves finding upper and lower bounds for the value of a summation, typically to understand the growth rate or behavior of a series. |
| Bounding with Analytic Methods | Bounding with analytic methods involves using mathematical techniques such as calculus, inequalities, or asymptotic analysis to establish upper and lower bounds for functions or expressions. |
| Bounding with Convex Hulls | Bounding with convex hulls involves utilizing the outermost convex polygon surrounding a set of points to establish upper and lower bounds for geometric quantities or optimize algorithms. |
| Bounding with Convexity | Bounding with convexity entails utilizing the properties of convex sets and functions to establish upper and lower bounds for quantities or optimize algorithms. |
| Bounding with Integrals of Step Functions | Bounding with integrals of step functions involves using the area under a piece wise constant function to approximate or bound the value of a more complex function. |

| | |
|---|---|
| Bounding with Mean Value Theorem | Bounding with the Mean Value Theorem involves utilizing the average rate of change of a function over an interval to establish upper and lower bounds for the function itself within that interval. |
| Bounding with Partial Sums | Bounding with partial sums involves estimating the sum of a sequence by considering only a finite number of terms, often used to approximate the behavior of infinite series or sequences. |
| Bounding with Probabilistic Methods | Bounding with probabilistic methods involves using probability theory to establish upper and lower bounds on the behavior of algorithms, data structures, or mathematical expressions, often based on random sampling or distributional properties. |
| Bounding with Symmetry | Bounding with symmetry entails leveraging symmetrical properties of objects or functions to establish upper and lower bounds for quantities, optimize algorithms, or simplify mathematical problems. |
| Bounding with Taylor Series | Bounding with Taylor series involves approximating a function using its Taylor series expansion and using the properties of the series to establish upper and lower bounds for the function within a given interval. |
| Breadth First Search | Breadth First Search is an algorithm used to traverse or search tree or graph data structures by exploring all neighbor nodes at the present depth before moving on to nodes at the next depth level. |

| | |
|---|---|
| Cardinality | Cardinality refers to the measure of the number of elements in a set, determining its size or quantity. |
| Cauchy Condensation Test | The Cauchy condensation test is a method used in mathematical analysis to determine convergence or divergence of a series involving non-negative terms by comparing it with a condensed series. |
| Child | A child in a tree refers to a node that is directly connected to another node, known as its parent, in a hierarchical tree data structure. |
| Chromatic Number | The chromatic number of a graph is the minimum number of colors needed to color the vertices of the graph such that no two adjacent vertices share the same color. |
| Closure | Closure in a relation refers to the property where if two elements are related, their relationship implies that they are still related even after applying a certain operation or transformation. |
| Co-domain | The co-domain of a function is the set that contains all possible output values or elements that the function can produce. |
| Combination | A combination is a selection of items from a larger set where the order of selection does not matter. |
| Commutative Law | The commutative law states that the order of operands can be changed without affecting the result of a binary operation. |
| Commutative Property | The commutative property states that the order in which two numbers are added or multiplied does not change the result. |

| | |
|---|---|
| Comparison Test | The comparison test is a method in mathematical analysis used to determine the convergence or divergence of a series by comparing it with another series whose convergence behavior is known. |
| Complement Law | The complement law in logic states that a statement and its negation cannot both be true simultaneously. |
| Complete Graph | A complete graph is a graph in which every pair of distinct vertices is connected by a unique edge, resulting in a graph where each vertex is adjacent to all other vertices. |
| Composite Methods | Composite methods in approximation by integrals involve dividing the interval of integration into sub intervals and applying numerical integration techniques such as the trapezoidal rule or Simpson's rule to each sub interval to approximate the integral over the entire interval. |
| Composition of Function | Composition of functions is a mathematical operation where the output of one function becomes the input of another function, resulting in a new function that represents the sequential application of the two original functions. |
| Compound Proposition | A compound proposition is a statement formed by combining simpler propositions using logical connectives such as "and," "or," "not," or "if-then." |

| | |
|---|---|
| Conditional | A conditional statement is a logical statement that asserts a relationship between an antecedent (the "if" part) and a consequent (the "then" part), stating that if the antecedent is true, then the consequent must also be true. |
| Conjunction | Conjunction is a logical operation that connects two propositions with the implication that both propositions must be true for the conjunction to be true, often represented by the word "and" or the symbol $\wedge$. |
| Connected Graph | A connected graph is a graph in which there exists a path between every pair of vertices, ensuring that no vertices are isolated or disconnected from the rest of the graph. |
| Constant Coefficient Recurrence Relation | A constant coefficient recurrence relation is a linear recurrence relation in which the coefficients of the terms do not depend on the index of the sequence, often encountered in the analysis of algorithms. |
| Contradiction | In logical equivalence, a contradiction arises when a statement and its negation are both true under the same conditions, resulting in the statement being logically false. |
| Converges | Converges refers to the property of a sequence or series where the terms approach a specific value as the index approaches infinity. |
| Countability | Countability refers to the property of a set being either finite or having a one-to-one correspondence with the set of natural numbers, allowing for its elements to be enumerated or listed in a systematic manner. |

| | |
|---|---|
| Counting | Counting in discrete structures involves determining the number of elements in a finite set or the number of possible outcomes in a finite or countably infinite scenario, often using combinatorial techniques. |
| Cut Edge | A cut edge, also known as a bridge, is an edge in a graph whose removal increases the number of connected components in the graph. |
| Cut Set | A cut set is a set of edges in a graph whose removal increases the number of connected components in the graph. |
| Cut Vertex | A cut vertex, also known as an articulation point, is a vertex in a graph whose removal increases the number of connected components in the graph. |
| Cyclic Graph | A cyclic graph is a graph that contains at least one cycle, which is a closed path of edges that starts and ends at the same vertex. |
| De Morgans Law | De Morgan's Law states that the negation of a conjunction (AND) is equivalent to the disjunction (OR) of the negations of the individual propositions, and the negation of a disjunction is equivalent to the conjunction of the negations of the individual propositions. |
| Degree of Vertex | The degree of a vertex in a graph is the number of edges incident to that vertex, indicating the number of connections it has with other vertices in the graph. |

**Depth**

Depth of a tree refers to the length of the longest path from the root node to a leaf node, representing the maximum number of edges that must be traversed to reach a leaf from the root.

**Depth First Search**

Depth First Search is an algorithm used to traverse or search tree or graph data structures by exploring as far as possible along each branch before backtracking, often implemented recursively or using a stack.

**Directed Graph**

A directed graph, also known as a digraph, is a graph in which edges have a direction, indicating a one-way relationship between vertices, allowing for directed paths and cycles.

**Disjunction**

Disjunction is a logical operation that connects two propositions with the implication that at least one of the propositions must be true, often represented by the word "or" or the symbol $\vee$.

**Distributive Law**

In logic, the distributive law states that a logical operation distributes over another operation, such as conjunction distributing over disjunction or disjunction distributing over conjunction.

**Diverges**

Diverges refers to the property of a sequence or series where the terms do not approach a specific value or the sum of the series does not converge to a finite value as the index or number of terms increases indefinitely.

**Domain**

The domain of a function is the set of all possible input values or elements for which the function is defined and produces an output.

| | |
|---|---|
| **Edge** | An edge is a connection between two vertices in a graph, representing a relationship or interaction between the corresponding elements. |
| **Edge Connectivity** | Edge connectivity in a graph refers to the minimum number of edges that must be removed to disconnect the graph, indicating the resilience of the graph to edge deletions. |
| **Equivalence Relation** | An equivalence relation is a relation on a set that is reflexive, symmetric, and transitive, allowing for the partitioning of the set into disjoint subsets called equivalence classes. |
| **Equivalences** | In logic, equivalences are statements or propositions that have the same truth value under all possible interpretations or conditions. |
| **Euler Graph** | An Euler graph is a graph that contains a cycle that includes every edge exactly once, visiting each vertex exactly once, known as an Eulerian cycle. |
| **Euler Path** | An Euler path is a path in a graph that visits every edge exactly once, though not necessarily every vertex, providing a traversal through the graph with a distinct starting and ending vertex. |
| **Finite Set** | A finite set is a set that contains a countable number of elements, meaning the elements can be enumerated or listed in a finite sequence. |
| **First Order Recurrence Relation** | A first-order recurrence relation is a linear recurrence relation where each term of the sequence is expressed as a linear combination of the preceding term and a constant coefficient. |

| | |
|---|---|
| Forest | In trees, a forest is a disjoint union of one or more trees, where each tree component does not contain any cycles. |
| Functions | Functions are mathematical relations that assign exactly one output value to each input value, representing a mapping from one set to another. |
| Gaussian Quadrature | Gaussian quadrature is a numerical integration technique that approximates definite integrals by choosing specific nodes and weights to accurately evaluate the integral using polynomial interpolation. |
| Generating Function | A generating function is a formal power series used in combinatorics and number theory to represent a sequence of coefficients or values in a compact and systematic manner, often facilitating the analysis of combinatorial problems. |
| Geometric Progression | A geometric progression is a sequence of numbers where each term after the first is found by multiplying the previous term by a fixed, non-zero number called the common ratio. |
| Geometric Series Bounding | Geometric series bounding involves estimating the sum of a geometric series by finding upper and lower bounds for its value, often using inequalities or other mathematical techniques. |
| Graph | A graph is a mathematical structure consisting of vertices (or nodes) connected by edges (or links), representing relationships or interactions between the vertices. |

| | |
|---|---|
| Graph Coloring | Graph coloring is a process of assigning colors to the vertices of a graph in such a way that no two adjacent vertices share the same color, commonly used in scheduling, map coloring, and various optimization problems. |
| Graph Isomorphism | Graph isomorphism is a concept in graph theory where two graphs are considered isomorphic if they have the same structure, meaning their vertices and edges can be matched in a one-to-one correspondence preserving adjacency. |
| Graph Theory | Graph theory is a branch of mathematics that studies graphs, which are mathematical structures used to model pairwise relations between objects, with applications in computer science, operations research, and various other fields. |
| Growth of Function | The growth of a function describes how its output values increase as its input values become larger, often analyzed in terms of asymptotic behavior or complexity. |
| Hamiltonian Graph | A Hamiltonian graph is a graph that contains a Hamiltonian cycle, which is a cycle that visits every vertex exactly once, except for the starting vertex, which is also the ending vertex. |
| Hamiltonian Path | A Hamiltonian path is a path in a graph that visits every vertex exactly once, providing a traversal through the graph with no repeated vertices. |
| Height | The height of a tree is the maximum number of edges on the longest path from the root node to any leaf node in the tree. |

| | |
|---|---|
| Homogeneous Recurrence Relation | A homogeneous recurrence relation is a linear recurrence relation in which each term is expressed as a linear combination of previous terms with constant coefficients and the right-hand side is zero. |
| Idempotent Law | In logic, the idempotent law states that applying a logical operation to a proposition twice yields the same result as applying it once, often represented as $P \wedge P = P$ or $P \vee P = P$. |
| Identity Law | In logic, the identity law states that a logical operation applied to a proposition combined with the identity element for that operation yields the original proposition, often represented as $P \wedge \text{True} = P$ or $P \vee \text{False} = P$. |
| Image | In function, the image refers to the set of all possible output values or elements that the function can produce when the domain elements are mapped through the function. |
| Incidence Matrix | An incidence matrix is a matrix representation of a graph where rows correspond to vertices and columns correspond to edges, with entries indicating whether a vertex is incident to an edge, commonly used in graph theory and network analysis. |
| Inequalities | Inequalities are mathematical expressions that compare two quantities, indicating that one is less than, greater than, or not equal to the other. |
| Inference Theory | In logic, inference theory pertains to the study of how logical statements or propositions can be derived or deduced from given premises using valid reasoning rules or inference techniques. |

| | |
|---|---|
| Infinite Set | An infinite set is a set that contains an unlimited or uncountable number of elements, extending indefinitely without bound. |
| Injective | Injective, also known as one-to-one, describes a function in which each distinct element in the domain maps to a distinct element in the codomain, without any two distinct domain elements mapping to the same codomain element. |
| Integral Approximation | Integral approximation involves estimating the value of a definite integral by using numerical methods to approximate the area under a curve, typically by dividing the interval of integration into smaller subintervals and approximating each subinterval's contribution to the total area. |
| Inverse Function | An inverse function is a function that undoes the action of another function, mapping the output values of the original function back to their corresponding input values. |
| Inverse Image | In a function, the inverse image refers to the set of all input values or elements that map to a specific output value or element under the function. |
| Irreflexive Relation | An irreflexive relation is a relation on a set where no element is related to itself, meaning there are no reflexive pairs in the relation. |
| Kruskals Algorithm | Kruskal's algorithm is a greedy algorithm used to find the minimum spanning tree of a connected, weighted graph by iteratively adding edges with the smallest weight that do not form a cycle. |

| | |
|---|---|
| Laws of Logic | The laws of logic are fundamental principles or rules that govern the behavior and relationships between propositions, guiding valid reasoning and inference in mathematics, philosophy, and computer science. |
| Leaf Node | A leaf node, also known as a terminal node, is a node in a tree data structure that does not have any child nodes, representing the endpoints of branches in the tree. |
| Limit Comparison Test | The limit comparison test is a method in mathematical analysis used to determine the convergence or divergence of a series by comparing it with another series whose convergence behavior is known, taking the limit of the ratio of their terms. |
| Linear Recurrence Relation | A linear recurrence relation is a mathematical relationship between the terms of a sequence, where each term is a linear combination of the preceding terms with constant coefficients. |
| Logical Connective | A logical connective is a symbol or word used to join or combine propositions in logic, determining the truth value of the compound statement formed by those propositions. |
| Loop | A loop in a graph refers to an edge that connects a vertex to itself, forming a cycle of length one. |
| Lower Bound | A lower bound is a value that serves as a limit from below, ensuring that the actual value of a quantity does not fall below this bound. |

**Majorization and Minorization**

In bounding summations, majorization and minorization techniques involve comparing a given series with another series that is easier to analyze or whose convergence properties are known, allowing for the establishment of upper and lower bounds for the original series.

**Master Theorem**

The Master Theorem is a mathematical tool used to analyze the time complexity of recursive algorithms by providing a concise formula for the runtime based on the structure of the recursive calls.

**Mathematical Induction**

Mathematical induction is a method of mathematical proof used to establish that a statement holds for all natural numbers by first proving it holds for the smallest natural number (usually 0 or 1), and then demonstrating that if it holds for an arbitrary natural number $n$, it must also hold for $n+1$.

**Minimum Spanning Tree**

A minimum spanning tree of a connected, undirected graph is a subset of its edges that connects all vertices together with the minimum possible total edge weight.

**Multi Graph**

A multigraph is a graph that allows multiple edges between pairs of vertices, possibly with different weights or labels on each edge.

**Negation**

Negation in the laws of logic refers to the logical operation that reverses the truth value of a proposition, typically transforming a true statement into false and vice versa.

| | |
|---|---|
| Node | In a tree structure, a node is a fundamental element that represents a single point of the overall structure, often containing data and linked to other nodes. |
| Non Linear Recurrence Relation | A non-linear recurrence relation is a mathematical relationship where each term depends on previous terms in a non-linear manner, often involving multiplication, exponentiation, or other non-linear operations. |
| Non-Homogeneous Recurrence Relation | A non-homogeneous recurrence relation is a mathematical relationship where each term depends on previous terms and is also influenced by a non-zero function of the index, known as the non-homogeneous or forcing term. |
| Numeric Function | A numeric function is a mathematical function that maps elements from a domain (often numbers) to elements in a codomain (also often numbers), yielding numerical values as outputs. |
| One to One Correspondence | One-to-one correspondence refers to a bijective relationship between the elements of two sets, where each element of one set uniquely matches with exactly one element of the other set and vice versa. |
| One to One Function | A one-to-one function, also known as an injective function, is a mathematical function in which each element of the domain maps to a unique element in the codomain, ensuring that no two different elements in the domain map to the same element in the codomain. |

| | |
|---|---|
| Onto Function | An onto function, also known as a surjective function, is a mathematical function where every element in the codomain is mapped to by at least one element from the domain, ensuring full coverage of the codomain by the function's range. |
| Parent | In a tree structure, a parent is a node that directly connects to and precedes one or more child nodes. |
| Partial Ordering | Partial ordering refers to a relation between elements in a set where some pairs of elements are comparable (related by the ordering relation), but not necessarily all pairs, and satisfies reflexivity, antisymmetry, and transitivity properties. |
| Permutation | A permutation is a rearrangement of elements of a set where each element appears exactly once, forming a specific order or sequence. |
| Pigeonhole Principle | The pigeonhole principle states that if more items are placed into fewer containers than there are containers, then at least one container must contain more than one item. |
| Planar Graph | A planar graph is a graph that can be embedded in a plane such that its edges intersect only at their endpoints, with no edges crossing each other in the plane. |
| Poset | A poset (partially ordered set) is a set equipped with a partial order relation that is reflexive, antisymmetric, and transitive, allowing for comparisons between elements based on the ordering relation. |

| | |
|---|---|
| Powerset | The powerset of a set is the set of all subsets of the original set, including the empty set and the set itself. |
| Predicate | In logic, a predicate is a statement that contains variables and describes a property or relation that may be true or false depending on the values assigned to those variables. |
| Prims Algorithm | Prim's algorithm is a greedy algorithm that finds a minimum spanning tree for a weighted undirected graph, starting from an arbitrary vertex and adding the shortest edge that connects a vertex not yet in the tree until all vertices are included. |
| Principle of Inclusion Exclusion | The Principle of Inclusion-Exclusion is a counting technique used to find the size of the union of two or more sets by correcting for over-counting of elements that belong to multiple sets. |
| Product Rule | If there are m ways to do a specific task and for each of these ways of doing the first task, there are n ways to do the second task, then there are m x n ways to do the procedure. |
| Propositions | Propositions are statements or assertions that can be either true or false, but not both simultaneously. |
| Range | The range of a function is the set of all possible output values (or function values) that the function can produce from its input domain. |
| Ratio Test | The ratio test is a method used to determine the convergence of an infinite series by evaluating the limit of the ratio of consecutive terms in the series. |

| | |
|---|---|
| Rectangular Rule | The rectangular rule, also known as the midpoint rule or rectangle method, is a numerical integration technique that approximates the area under a curve by dividing it into equally spaced intervals and using the height of the function at the midpoint of each interval to compute the area of rectangles. |
| Recurrence Relation | A recurrence relation is a mathematical equation that defines a sequence where each term is defined in terms of one or more preceding terms in the sequence. |
| Recurrence Trees | Recurrence trees are graphical representations used to analyze and solve recurrence relations, particularly in the context of algorithm analysis and complexity determination. |
| Reflexive Closure | Reflexive closure refers to the addition of pairs to a relation such that every element is related to itself, ensuring reflexivity in the relation. |
| Reflexive Relation | A reflexive relation on a set is a binary relation where every element is related to itself; formally, for all elements x in the set, (x,x) belongs to the relation. |
| Regular Graph | A regular graph is a graph where each vertex has the same number of edges (degree), meaning all vertices have identical connectivity within the graph. |
| Relation | A relation is a set of ordered pairs that establishes a connection or association between elements from two or more sets. |

| | |
|---|---|
| Root | In a tree structure, the root is the topmost node that serves as the starting point and has no parent node. |
| Root Test | The root test is a convergence test used in series analysis that examines the limit of the n-th root of the absolute value of the terms in a series to determine its convergence behavior. |
| Rule CP | If we can derive 'S' from 'R' and a set of premises, then we can derive R S from the set of premises alone |
| Rule P | A premise may be introduced at any point in the derivation. |
| Rule T | A formula 'S' may be introduced in a derivation, if 'S' is tautologically implied by one or more of the preceding formulas in the derivation. |
| Second Order Recurrence Relation | A second-order recurrence relation is a mathematical relationship where each term depends on its two preceding terms in the sequence. |
| Sequence | In generating functions, a sequence refers to a series of coefficients representing terms of a discrete mathematical structure, typically arranged in a formal power series to facilitate analysis and computation of properties such as sums and recurrence relations. |
| Simple Graph | A simple graph is an undirected graph where no two edges connect the same pair of vertices (no multiple edges) and there are no loops (edges connecting a vertex to itself). |

**Simpson's Rule** — Simpson's rule is a numerical integration method that approximates the area under a curve by fitting parabolic segments between points and using these segments to compute the integral.

**Spanning Tree** — A spanning tree of a connected graph is a subgraph that includes all vertices of the original graph and is a tree (acyclic connected graph).

**Subset** — A subset is a collection of elements that are all contained within another set, including the possibility of being equal to the original set itself.

**Substitution Method** — The substitution method is a technique used in solving recurrence relations by assuming a form for the solution and then proving its correctness through substitution and mathematical induction.

**Subtree** — A subtree in a tree structure is a subset of nodes and edges that form a tree within the original tree, rooted at some node of the original tree.

**Sum Rule** — If a task can be done in either one of m ways or one of n ways and none of the set of m ways is same as the set of n ways, then there are $m + n$ ways to do the task.

**Summation Formulas** — Summation formulas refer to mathematical expressions that provide a concise way to compute the sum of a series of numbers or terms, often involving specific patterns or sequences.

**Summation Properties** — Summation properties are rules and identities that govern the manipulation and evaluation of sums, including properties related to constants, variables, and operations such as addition, multiplication, and distribution.

| | |
|---|---|
| Superset | A superset is a set that contains all the elements of another specified set, possibly along with additional elements. |
| Surjective | Surjective, or onto, describes a function where every element in the codomain is mapped to by at least one element in the domain, ensuring full coverage of the codomain by the function's range. |
| Symmetric Closure | The symmetric closure of a relation includes all pairs that are symmetric with respect to the original relation, ensuring that for every pair (a,b) in the relation, (b,a) is also included. |
| Symmetric Relation | A symmetric relation is a binary relation on a set where if (a,b) is in the relation, then (b,a) is also in the relation, meaning the relation is bidirectional between pairs of elements. |
| Tautology | A tautology is a statement in logic that is always true, regardless of the truth values of its components or variables. |
| Telescoping Series | A telescoping series is a series where most of the intermediate terms cancel out when the consecutive terms are subtracted from each other, leaving only the first and last terms to contribute to the sum. |
| Time Complexity | Time complexity refers to the computational efficiency of an algorithm in terms of the amount of time required for it to complete its task, typically measured as a function of the input size. |

| | |
|---|---|
| Transitive Closure | Transitive closure refers to the smallest transitive relation that contains a given relation, ensuring that if a is related to b and b is related to c, then a is related to c as well. |
| Transitive Relation | A transitive relation on a set is a binary relation where if a is related to b and b is related to c, then a is also related to c, ensuring the relation propagates through the elements in a chaining manner. |
| Trapezoidal Rule | The trapezoidal rule is a numerical integration method that approximates the area under a curve by using trapezoids whose bases lie along the x-axis and whose tops connect to the curve. |
| Traversal | Traversal refers to the systematic visiting or exploration of each node or element in a data structure, graph, or tree, often following a specified order or pattern. |
| Tree | In graph theory, a tree is an acyclic connected graph where there is exactly one path between any pair of vertices, ensuring it forms a connected and minimal subgraph. |
| Truth Table | A truth table is a systematic way of representing the truth values of logical expressions by listing all possible combinations of truth values for the variables involved and showing the resulting truth value of the expression for each combination. |
| Union of Set | The union of two sets is the set containing all elements that are in at least one of the original sets, denoted by $A \cup B$. |

| | |
|---|---|
| Upper Bound | In time complexity analysis, an upper bound refers to the worst-case scenario for the amount of time an algorithm requires to solve a problem, often denoted using Big-O notation. |
| Vertex | In graph theory, a vertex (plural: vertices) is a fundamental unit representing a point or node within a graph, which may have connections to other vertices via edges. |
| Vertex Connectivity | Vertex connectivity in graph theory refers to the minimum number of vertices that need to be removed in order to disconnect the graph or make it trivial (consisting of isolated vertices). |
| Weighted Graph | A weighted graph is a graph where each edge is assigned a numerical weight or value, typically representing a cost, distance, or some other metric associated with traveling between the vertices connected by that edge. |
| Well Formed Formula | A well-formed formula (WFF) is a syntactically correct arrangement of symbols and operators in a formal language or logic system, adhering to the grammar rules of the system. |

# REFERENCES

**Books**

1. Kenneth H. Rosen, "Discrete Mathematics and Its Applications", McGraw-Hill Education, 2018.

2. Susanna S. Epp, "Discrete Mathematics with Applications", Cengage Learning, 2018.

3. Ralph P. Grimaldi, "Discrete and Combinatorial Mathematics: An Applied Introduction", Pearson, 2018.

4. Richard Johnsonbaugh, "Discrete Mathematics", Pearson, 2017.

5. Norman L. Biggs, "Discrete Mathematics", Oxford University Press, 2002.

**Websites**

1. https://www.javatpoint.com/discrete-mathematics-tutorial

2. https://archive.nptel.ac.in/courses/106/105/106105192/

3. https://www.geeksforgeeks.org/discrete-mathematics-tutorial/

4. https://www.tutorialspoint.com/discrete_mathematics/index.htm

5. https://learn.saylor.org/course/CS202

## K

Kruskals Algorithm 260

## L

Laws of Logic 236, 261
Leaf Node 261
Limit Comparison Test 76, 261
Linear Recurrence Relation 261, 263
Logical Connective 261
Loop 137, 261
Lower Bound 79, 261

## M

Majorization and Minorization 76, 262
Master Theorem xii, 131, 262
Mathematical Induction xi, 57, 262
Minimum Spanning Tree 195, 262
Multi Graph 262

## N

Negation 211, 212, 217, 222, 232, 237, 262
Node 191, 261, 263
Non-Homogeneous Recurrence Relation 263
Non Linear Recurrence Relation 263
Numeric Function 263

## O

One to One Correspondence 16, 263
One to One Function 13, 263
Onto Function 15, 264

## P

Parent 189, 193, 264
Partial Ordering xi, 36, 37, 264
Permutation xi, 46, 53, 70, 264
Pigeonhole Principle xi, 43, 44, 264
Planar Graph xii, 147, 174, 176, 182, 264
Poset 264
Powerset 265
Predicate 207, 265
Prims Algorithm 265
Principle of Inclusion Exclusion xi, 60, 265
Product Rule 40, 265
Propositions xii, 205, 206, 207, 208, 265

## R

Range 10, 11, 23, 89, 90, 265
Ratio Test 77, 265
Rectangular Rule 81, 266
Recurrence Relation 126, 253, 256, 259, 261, 263, 266, 267
Recurrence Trees xii, 125, 266
Reflexive Closure 32, 266
Reflexive Relation 26, 266
Regular Graph 142, 266
Relation 26, 27, 28, 29, 34, 37, 38, 39, 126, 247, 253, 256, 259, 260, 261, 263, 266, 267, 269, 270
Root 77, 189, 267
Root Test 77, 267
Rule CP 241, 242, 243, 244, 267
Rule P 241, 243, 244, 267
Rule T 241, 243, 244, 267